T0311837

Dynamic Positioning for Engineers

Dynamic Positioning for Engineers

Surender Kumar

CRC Press
Taylor & Francis Group
Boca Raton London New York

CRC Press is an imprint of the
Taylor & Francis Group, an **informa** business

Contents

Foreword

by Capt. Rajesh Tandon

V Group Global Director – Industrial Relations and Seafarer Development
Past Chairman – International Maritime Employers Council, London

I am delighted to write this Foreword, not only because Dr. Surender Kumar worked with V Group as the principal of the Global Maritime Training Centre (previously known as the Sir Derek Bibby Maritime Training Centre), but also because I firmly believe in the educative value of this book for readers, especially those in the field of dynamic positioning (DP).

Dr. Kumar and his team were instrumental in introducing DP training in India. He earned the admiration of his students and that of the industry while imparting effective DP training skills. The DP laboratory he set up was impressive and proved to be an effective tool in learning and training. He has the distinction of having trained seafarers from more than 24 nationalities, and has rightfully, albeit informally, earned the title of "DP Guru". His sincere efforts earned him the best trainer award for offshore courses in 2018. It was during his extensive international DP training experience that he thought of the idea to write *Dynamic Positioning for Engineers.*

The book sufficiently covers everything a reader needs to know about dynamic positioning. The book starts with an introduction to the offshore industry and guides the reader through the basic concepts of DP, the six degrees of freedom, the seven components of a DP system, and the equipment class. The book is primarily written for the benefit of engineers and other DP professionals to gain better insight into DP systems. It covers the system perspective, thus making it easy to understand and be prepared for fault identification and tracing in various components like DP computers, environmental sensors, position reference sensors, thrusters and DP hardware.

The chapters on thruster automation, DP controls, various types of signals used in DP automation and power management systems make this book a must-read for all DP engineers and technical personnel in this field. The other important aspects, such as consequence analysis, capability plot, FMECA and DP documentation, offer something for everyone in the DP field. This book will not only provide an effective learning experience, but can also be used as a reference resource for dynamic positioning.

I passionately believe that Dr. Kumar will receive very constructive and grateful feedback from readers across the world. I take this opportunity to thank him for inviting me to write the Foreword and compliment him for this initiative.

Foreword

by Venkatraman Sheshashayee

Former CEO, Greatship, Jaya Holdings and Miclyn Express Offshore
Current Managing Director, Radical Advice

It is my pleasure and privilege to write this Foreword for this very special book, *Dynamic Positioning for Engineers*, authored by Dr. Kumar.

I have known Dr. Kumar from the days when he was the Principal of the Sir Derek Bibby Maritime Training Centre in Mumbai. His team pioneered the introduction of DP training in India. He has more than a decade's experience teaching maritime professionals in India, Nigeria, South Africa, Dubai, Korea Maritime and Oceans University, Busan, South Korea and Indonesia, earning justified popularity amongst his students. Dr. Kumar's vast experience teaching DP to international students inspired him to write this book, catering to the needs of the technical staff and others who intend to learn the technical aspects of dynamic positioning.

Having gone through the contents of the book carefully, I feel confident that it is both elegant and educational. It is a tailormade reference manual for engineers working onboard DP-enabled vessels. The introduction to offshore and the history and development of dynamic positioning are well covered. The basic concepts of dynamic positioning explained in chapters like six degrees of freedom, the seven components of DP, equipment class and different types of DP vessels will make the readers relate to the subject with ease. The concepts of electrical propulsion, thruster controls and automation, power management system used on DP-enabled vessels, working principles of environmental sensors and position reference sensors, signals used in interfacing the sensors and thrusters with DP systems are explained in easy-to-understand language.

The author has emphasised the need for, and the importance of, documentation used and maintained on board DP vessels, tests and trials like FMECA, consequence analysis and capability plots.

I sincerely hope that Dr. Kumar's efforts will be recognised and appreciated by the international offshore community and that the book will prove to be a good learning tool. It has all it takes to be the seminal reference guide to dynamic positioning.

Happy reading!!

Preface

Learning and teaching DP for about two decades gave me the idea to write this book, *Dynamic Positioning for Engineers*. My interaction with thousands of students in India and abroad, DP professionals, experts in the field, DP and sensor manufacturers and the Nautical Institute, London, motivated me to write it. Student feedback after training revealed that my methods of explanation and teaching were amazingly simple and effective, and this became a catalyst for me to start writing this book. It is important for every DP professional to understand well the system that they are going to operate or maintain. Neither operating nor maintaining is an easy task without knowing those systems well.

This book starts with an introduction to the offshore industry, the need for and the development of DP systems, making the reader aware of historical developments. Further chapters, in sequence, introduce the reader to the DP system, and the subsystems and components of the DP system. The six degrees of movement and how they are measured, as well as what is a control loop to control the movements on the horizontal plane, are explained in very easy language.

The understanding of a system is incomplete without knowing the automation and system integration with various subsystems and components. The chapters on thruster automation, sensors and position reference sensors fulfill the requirements of understanding these aspects. Explaining the various signals associated with the thrusters, sensors and PRS, in a way that the reader can get to know the real issues behind the correct functioning, interfacing and networking of the system, is a unique feature of this book.

I take this opportunity to thank my teachers and students who taught me a lot. I thank my colleagues and team members at the Sir Derek Bibby Maritime Training Centre, Global Maritime Training Centre, Aquamarine Maritime Academy, Mumbai, Ocean's XV DP Training Centre, New Delhi, HIMT Offshore, Chennai, Maritronics DP Training, Dubai, Korea Maritime and Oceans University, Busan, Charkin DP Training Centre, Port Harcourt and Azureus Simulators Asia, Jakarta. I would also like to thank my parents, my wife Rajeshwari and my kids Rahul and Susmita, as without their contributions and perseverance, this book would not have become a reality.

Dr. Surender Kumar
(CEng, CMarEng, CMarTech, FIMarEST, FNI, fDPO, PhD)

Author

Dr. Surender Kumar is an electrical engineer by profession, having years of experience with different kinds of ships including naval, mercantile fleet and offshore vessels as an ETO/electrical officer. He also has managerial experience ashore prior to taking up maritime training and it is his full-time interest. An academician in his own right, Dr. Kumar passed his national eligibility test (NET) from the University Grants Commission to teach management subjects in 1996. Ever since, he has been a visiting faculty member for management programmes for various prominent institutes in Mumbai and Bangalore. He is also an approved guide for PhD scholars.

His thesis entitled, "Effectiveness of Seafarer's Training Using Maritime Simulators" from the University of Petroleum and Energy Studies, Dehradun, India, earned him his PhD. He is an internationally experienced maritime trainer genuinely interested in learning and teaching dynamic positioning.

Dr. Kumar was the principal of an internationally acclaimed training centre. He has been instrumental in introducing DP training to India and has also been responsible for introducing MCA- (UK) approved training for the first time in India. He has enriching experience in setting up DP training facilities in India, South Africa, Nigeria and Indonesia.

His other qualifications and associations are as follows:

- Life Member – All India Management Association, New Delhi
- Fellow – Institute of Marine Engineering, Science and Technology, London (FIMarEST)
- Fellow – The Nautical Institute, London (FNI)
- Chartered Marine Engineer – UK (CMarEng)
- Chartered Engineer – UK (CEng)
- Chartered Marine Technologist – UK (CMarTech)

1 Introduction to Offshore

The offshore industry, which is also known as the oil and gas industry, is considered one of the most technologically advanced industries in the world. The oil and gas industry has played a vital role in modern civilisation and will probably influence it in the future as well, until practical alternatives are found and proved.

There is good historical evidence which indicates that even in Noah's time, boats were repaired and coated with tar for stopping leakages. The earliest use of tar in history is mentioned in the ancient Indian civilisation of Mehrgarh for storage of grains somewhere around 4000 BC. As time progressed, mankind found various other uses for this wonderful product which influences our lives greatly. So much so, that if we seriously look at the daily use of products around us, more than 80% of them have an input from hydrocarbons.

The oil and gas industry is popularly also known as the hydrocarbon industry. Hydrocarbons, as defined by Schlumberger (n. [Geology]) is, "A naturally occurring organic compound comprising hydrogen and carbon". How many atoms of C and H are combined together to form a molecule of hydrocarbon defines the final product. It may be a simple combination like methane (CH4 – it has one atom of carbon and four atoms of hydrogen) and can go on to form a very complex molecule to form a gas, liquid or a solid like coal and tar.

Hydrocarbon exploration, which is also referred to as oil and gas exploration, may be defined as an activity by a team of experts to plan, search and collect the hydrocarbon deposits under the Earth's surface. Hydrocarbons may include oil and natural gas.

Being a resource-extensive activity, hydrocarbon exploration is generally a state-controlled activity or is owned by oil majors in this industry. There is a big difference in the exploration of hydrocarbons onshore and offshore. As we enter offshore fields, every activity becomes much more expensive. Safety also demands higher levels of investment.

Historically, oil was first discovered naturally in China by a nomadic group when they saw smoke emanating from the nearby hills in the winters. No one really knew what the black material oozing out was!! In the beginning of the eighteenth century a well was drilled about 30 m off the coastline near Baku. It was not a successful venture, but it was marked as the beginning of a new era. The efforts finally led to successfully producing oil in the Caspian Sea in 1925. In India, once again, oil was found by chance in the jungles of Assam by a British Indian Army officer and this led to the first oil being struck, in 1867. Thus, India also ranks among the first to have engaged in oil exploration.

Initially, the oil and gas industry was restricted to onshore locations. It was later that attention was drawn to the resource-rich coastal areas. Technically, oil and gas operations offshore are considered more challenging than those which are land-based. There is a big influence on these by the prevalent environment. Safety of operations is and has always been a major challenge. Large offshore production facilities will need large investments; hence, the stakes are very high.

The Americans came to the forefront due to technological advantage and took the lead at this stage. The Californian oil wells slowly moved into the sea. These oil wells were connected by piers made of wooden (bamboo) structures in those days and were slowly replaced by steel structures.

Offshore drilling started in 1869 from a jack-up rig but when drilling in deeper water it was not possible to install jack-up rigs or use anchors any more due to depth restrictions, strength of the material used and, more importantly, the safety issues.

Thus, the need for an alternative method was felt and it was Howard Shatto, an enterprising engineer working for Shell, who came up with a basic idea which later took the shape of the modern dynamic positioning (DP) system. This is how DP was born, and the first DP was used in 1961 on board a vessel called *Eureka* working for a Shell project carrying out drilling at a depth of 300 m. This was followed by another vessel, *Calldrill*, in 1964, used for drilling in depths of 2000 m. Since then there have been lots of technological advancements and hence the depth and safety of operations have been significantly improved using DP-enabled vessels. Though the initial use of DP was for drilling operations, as time passed more applications and vessels started making use of this technology. Today the use of DP has spread over not only the entire offshore activities, but also to the main fleet and other activities.

Exploration activities for oil and gas are considered expensive and these involve a certain degree of risk. Considering the harsher environment, nature of investment made, offshore exploration activities are considered more expensive and riskier business activities. Generally, offshore exploration activities are undertaken by very large corporations or government agencies. For example, a normal shallow water oil well in the North Sea may cost tens of millions of Euros but the cost of a deep water well may go up to hundreds of millions of Euros, making it a really expensive business.

1.1 DRILLING OPERATIONS

Well intervention and drilling are considered the biggest and the most critical investment for any oil company. The collection of all the required data and analysis is considered important, so as to bring the learning curve and expense down. Thereafter, the drilling company follows a well-documented and safe programme to complete the drilling activities.

In general, oil well activities may be divided into the following stages:

- Planning stage.
- Drilling stage.
- Completion stage.
- Production stage.
- Abandonment stage.

After the planning as mentioned above, drilling activity is executed. A well is drilled into the earth and the diameter of the hole is generally between 13 and 76 cm using a drill bit from a rig. After drilling a hole, a steel pipe, normally referred to as "casing", having a diameter slightly smaller than the hole, is inserted in the hole. Using cementing operations, the pipe (casing) is now secured. This provides strength to the drilling hole. A production tube with a diameter a little less than the casing is used as the production pipe.

Once the drilling casing is in place the well is considered to be "completed". Well completion may be defined as the process by which the well is prepared to produce oil or gas. The casing is then perforated by drilling holes by using explosives in the depths and areas called the "production zone". The perforation or the holes in the pipe will provide the passage for the oil trapped in the source rock to flow into the production pipe.

The most important stage in this process is the production stage. This is the stage where the real production of oil and gas is achieved. The well may have been installed with a "Christmas tree", which is a group of valves. These valves are utilised to regulate pressure and flow. They also provide access to the well bore if further completion work is necessary. From here the flow may be controlled for the purposes of distribution.

It is also very common to classify oil wells by the purpose they serve. Thus, they can be classified as:

- "Wildcat wells", as the name suggests, are drilled with the hope of finding oil. Generally, not much is known about the subsea surface or subsurface prior to drilling. Industry experts may also use this as an appraisal well. In a true sense an appraisal well is used to assess various characteristics such as the flow rate of a reservoir which has been in use.
- "Exploration wells" are those wells which are mainly drilled for data collections.
- "Production wells" are important as production is from these types of wells, which are active after the data collection stage and it is found that they meets the requirements of production.

1.2   OFFSHORE STRUCTURES

To carry out various oil-related activities at sea, offshore structures are used. A very commonly seen structure is an offshore platform. This structure provides all the required support and it facilitates installation of the required equipment which is used for the main/auxiliary activities.

Whether drilling onshore or offshore, the first activity undertaken is known as the "prospecting phase". To start this process, depending on the water depth and other requirements, either a floating rig or a jack-up barge may be used. An exploration well is drilled to find out if there is a possibility of finding hydrocarbon products in the location (Figure 1.1).

Depending upon the viability, the oil major company may decide to make use of a permanent production platform to continue the process to extract the oil and gas resources. This stage is called the "production stage".

FIGURE 1.1   Jack-up rig.

The different types of offshore structures used for different purposes are given below:

- Oil and gas exploration.
- Oil and gas production.
- Processing platforms.
- Accommodation for workforce.
- Bridges and walkways/causeways.
- Offshore loading and off-loading facilities.

1.2.1 Fixed/Stationary Platforms

It is common to find such structures offshore where these steel structures are fixed to the bottom of the sea using methods like piling done at the seabed. Sometimes these structures are concrete structures which are directly placed at the seabed using gravity. (Figure 1.2).

FIGURE 1.2   Drilling rig.

1.2.2 JACKET PLATFORMS

Jacket platforms are structures using a number of legs depending on the design. The legs are generally connected to the structure using tubular bracing members. The legs are supported by piling done deep into the seabed (Figure 1.3).

1.2.3 STAR PLATFORMS

In the North Sea and other parts of the world there are also structures used which are commonly known as a Slim Tripod Adapted for Rigs (STAR).

1.2.4 COMPLIANT TOWERS

Compliant towers are offshore structures which consist of a flexible tower which is connected and supported by a piled foundation. The tower generally supports structures used for drilling and production operations. Compliant tower structures can withstand significant forces and are generally used up to depths 500 metres.

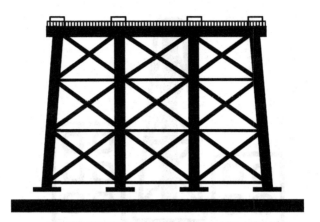

FIGURE 1.3   Jacket platform.

1.2.5   Semi-Submersible Platforms

These structures/platforms are constructed on the same principle as that of marine vessels, wherein sufficient buoyancy is ensured to keep them afloat and stabilised. These structures are manoeuvrable and have facilities which can be ballasted/deballasted. These structures may be moored/anchored, or in many cases may use a DP system to keep themselves in a desired position.

Although there are no standard guidelines, these structures may be used within depths of about 180 to 1800 metres.

1.2.6   Single Point Anchor Reservoir (SPAR) Platforms

There is a big similarity between a tension-leg platform (TLP) and a spar platform, i.e., they are both moored to the seabed. A TLP uses vertical tensioners but the spar uses normal mooring lines. Industry practice shows that the spars have three design configurations as below:

- One-piece cylindrical hull type generally referred to as "conventional".
- Truss spar wherein the middle section is made up of truss elements.
- Cell spars which are made up of many vertical cylinders.

Spars may be preferred over TLPs, especially for the small size reservoirs.

1.2.7   Tension-Leg Platforms (TLPs)

TLPs are platforms consisting of floating rigs which are fixed to the seabed by the use of pre-tensioned tethers. TLPs are preferred for use in water depths of about 2000 metres. Structurally they may appear similar to the semisubmersibles.

1.2.8 Jack-Up Platforms

The situations and environmental conditions may demand that the structure may have to be raised above the sea surface to avoid the effect of environmental forces. These structures are generally used in shallow waters and are designed so that they can be easily moved from one field to another.

1.2.9 Floating Production Systems

At times, situations may demand use of larger vessels or ships equipped with processing facilities, and these floating production systems are generally referred to as floating production storage and offloading (FPSO) (Figure 1.4).

If these units are being used simply as storage and offloading units, they are referred to as a floating storage and offloading system (FSO) or floating storage unit (FSU).

1.2.10 Subsea Production Systems

As oil and gas exploration extends into deeper waters, subsea systems are playing an important role. Subsea wells and other such facilities are a reality now and being used in many proven production fields.

Subsea production control systems play an important role in the offshore industry for the production of hydrocarbons. The system may consist of the following components:

- Mechanical.
- Electrical.

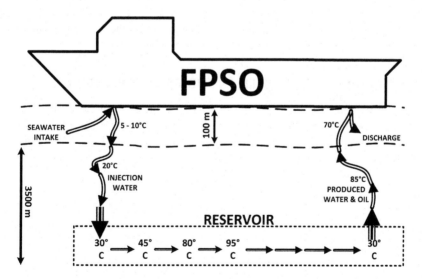

FIGURE 1.4   FPSO.

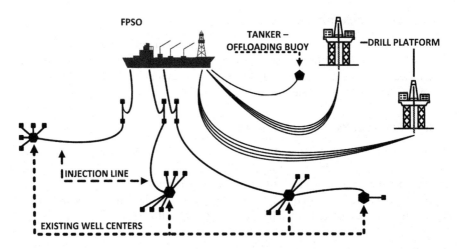

FIGURE 1.5   FPSO with other drilling units.

- Hardware.
- Software.
- Interface.
- Fibre optics.
- Instrumentation.
- Communication, etc.

Christmas tree, for example, provides a method to contain pressure for safely cap-
ping the oil well via the interface it provides with the subsea well head. There are
many subsystems put together to make this happen. The system may be configured
as an "open loop system" or a "closed loop system" (Figure 1.5).

1.3   CONTROL SYSTEM USED IN SUBSEA PRODUCTION

ISO 13628-6 defines a subsea control system as a "control system operating a subsea
production system during operation". Practically it may be defined as a system which
is designed for safe operations ensuring and maintaining the extraction of oil and gas
from facilities in the subsea environment.

2 Historical Background to Development of Dynamic Positioning

2.1 BACKGROUND AND HISTORY OF DYNAMIC POSITIONING DEVELOPMENT

It was in early 1960 that dynamic positioning (DP) as a new technique was first applied and used in carrying out drilling, coring, diver support or cable laying. Now it is more than five decades since the offshore industry started making use of this technology.

In the beginning, most of the DP-enabled vessels were converted vessels. The conversions of conventional ships were modified by fitting extra thrusters and computers to process information from the various sensors, including one to measure the position of the ship. Initially, the most utilised position sensor was the taut wire system.

The initial developments were that the vessel, a specialised vessel those days, with two rudders and two main propellers, was augmented with a bow thruster. This helped the vessel to hold its heading without the use of the main propellers and the rudders. As the thruster was installed in the bow, it was named "bow" thruster. This was followed by a thruster in the "stern". Using both the bow and the stern thrusters in combination could help manoeuvre the vessel easily (Figure 2.1).

This was followed by the addition of new devices on the bridge wherein all the thrusters were able to be controlled by an independent lever. This made the job easier for the bridge to control the vessel's position but was still heavily dependent on the expertise and experience of the person on watch. This led to the addition of more automation to this function, although marine automation was in its infancy that time.

A computing device to a "joystick", was now added. This joystick made it possible to manoeuvre the vessel by using just one lever rather than many levers depending on the number of thrusters. In those days this was a big development and helped to manoeuvre the vessel safely. Due to rapid development in the field of computers, the computing device was replaced with a much more versatile computer called a "real time computer" (RTC). This computer was given inputs from the environmental sensors and position reference sensors.

A vessel reference model, or mathematical model of the ship, was developed, and inputs given to the RTC. The RTC then had good enough information combined with the mathematical model. All this information was processed and an output known as thruster allocation logic (TAL) was the result. This TAL was able to control the

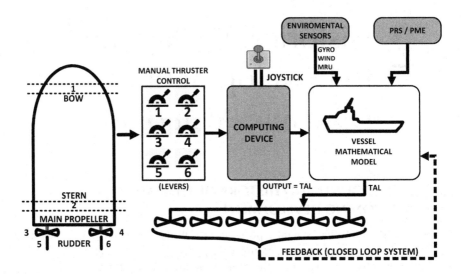

FIGURE 2.1   DP system development.

required number of thrusters. To make it a closed-loop system a feedback was connected from the thrusters back to the controller. Many refinements and technological improvements can now be seen in modern DP systems.

Back then, North Sea activities were moving into deeper waters and thus the need to have safer ways to drill in these waters was required. Deeper waters and severe weather conditions were probably the main reasons for developing DP systems on the vessels involved in offshore operations, mainly drilling, in those days.

Unfortunately, some incidents were observed on board the diving support vessels (DSVs) using DP systems which made the industry feel that some more safety measures had to be included to make these vessels more reliable. Initially the operators and the charterers were not happy with the DP system on board these vessels.

During the 1980s, big developments were observed, making the DP vessels safer and more reliable. The operations were moving into deeper waters and this required them to have redundancy in the DP system components. The need was felt to train the operators and the other technical staff associated with the DP system.

The safety requirements and hence the developments were felt to be more important for DSVs and the drilling ships. Once these developments were successfully implemented in these vessels, it was easy to apply them to the remaining offshore fleet vessels.

Dynamic positioning is one of the fastest-growing technologies. Although it was born out of the necessity of drilling safely in deep waters, it is now being used for a wide number of applications and its demand increased further as the concept became very popular in other shipping and offshore activities.

As the time and the technology progressed, by the late 1970s DP techniques were widely accepted for the purpose of making a vessel hold its position for offshore operations. By this time DP was an established and accepted technique in offshore vessel positioning. It is estimated that from a very humble number of approximately 65 DP-enabled vessels in 1980, the number today is around 3600.

It is worth noting here that the functions, use and diversity of DP vessels in the last 40 years or so have undergone tremendous changes. DP systems are being used not only in the offshore industry but also in sectors like cruise liners.

Howard Shatto, better known as the father of dynamic positioning, came up with the initial idea to control vessels' movements, and his efforts resulted in making the world's first dynamically positioned vessel a reality. Drilling operations were carried out by the vessel *Eureka*; she was modified to be controlled safely and satisfactorily for drilling operations.

Attempts were made to ensure that vessel's movements, namely the surge, sway and yaw controls were independently controlled. Force vectors were measured, and these vectors were used in design controls so that thrust could be generated using TAL and suitable thrust vectors were generated. This TAL was then assigned to the available thruster/s under DP control.

The International Marine Contractors Association (IMCA) defines a dynamically positioned vessel as a marine unit or a ship that has the capabilities to maintain its position at a fixed location, or may be moving on a predefined track by using active thrust. Active thrust is the thrust which is generated by the thrusters available on board.

3 The Seven Components of a DP System

A dynamic positioning (DP) system is comprised of the following seven components:

1. Power.
2. Thrusters.
3. Environmental sensors.
4. Position reference sensors /Position measuring equipment (PRS/PME).
5. Computer/real time computer (RTC)/DP controller (DPC)
6. MMI/HMI.
7. Operator.

3.1 SUBSYSTEMS AND GROUPING

For easy understanding, the seven components may be grouped as below:

a. Power and thruster group.
b. Sensors and PRS group.
c. Computer, HMI and operator group. (Figure 3.1).

3.1.1 POWER

The power in the following discussions is restricted to "electrical power", as the majority of the DP vessels on the market are electrical propulsion vessels. If the figures from the market are to be believed, the new vessels are invariably electrically propelled vessels.

Starting from nominal power consumption on board general offshore vessels, large drilling vessels may have onboard power systems up to 50 MW. The load demand is met by installing the appropriate number of generators running in parallel on a single bus or on isolated buses using a bus tie connector. The position and configuration of bus tie breakers would depend upon the class of DP vessel and the job requirements. Most often these generators make use of a diesel prime mover and hence the system is generally called a diesel electrical propulsion system.

To understand the outline or the single line diagram for a power system of any vessel DP vessel, it is important to know how many generators the ship has been installed with and how these are connected with the main switchboard, which is also referred to as "Bus" or "Busbar".

The two or three bus or switch boards of the ship may be connected via the bus tie breakers. These bus tie breakers are generally kept open assuring of safe operations.

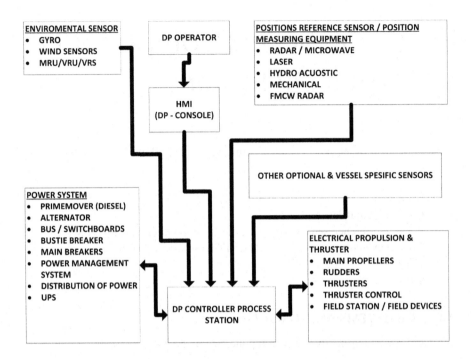

FIGURE 3.1   The seven components of a DP system.

It may be noted that grounding transformers are also installed for main switchboards. Each of the high voltage (HV) switchboards then supply power to bus supplies power to 480/440/380 V switchboards for ship services. The HV switchboards also may supply special tasks/services through special switch board/s. The 480/440/380 V switchboard then supply power to the 220/120 V small users and lighting services.

3.1.2 THRUSTERS

The thrusters use the electrical power generated by the power plant and help the vessel to maintain her position. This pair of thrusters and power systems ensures the position-keeping capability and thus can be treated as an inseparable pair.

A thrust-generating device which is capable of generating thrust in an athwartship direction and is fitted inside a purpose-built tunnel. These tunnels are designed in the bow or the stern of the ship, and the thrusters fitted inside these tunnels are accordingly called a bow thruster or a stern thruster (Figure 3.2).

A bow thruster can make manoeuvring and docking easier for conventional vessels. For DP-enabled vessels the bow thruster can be used for various purposes such as heading control, and sway control in association with stern thrusters and other propulsion devices.

A stern thruster also functions in the same way as the bow thruster. A bigger and more powerful DP vessel may have more than one bow/stern thruster to augment her position-keeping capability.

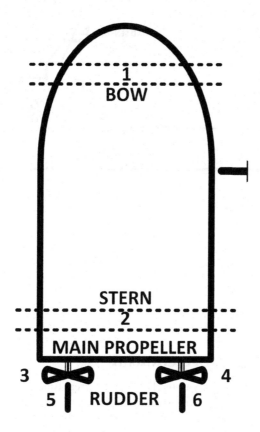

FIGURE 3.2   Thrusters (two).

3.1.3 ENVIRONMENTAL SENSORS

The sensor group includes the environmental sensors and the position reference sensors. The environmental sensors include wind sensors, vertical reference units (VRU) or vertical reference sensors (VRS), also known as a motion reference unit (MRU). A gyro is included in the environmental sensor group, as a gyro measures the results of the forces which are acting on the ship as a result of which the ship may change its heading, The rate of change is also measured by the gyro which is commonly known as rate of turn (ROT) (Figure 3.3).

Any heading change is detected by the gyro and fed to the DP system (controller) so that an appropriate counterthrust may be applied to change back to the desired heading.

3.1.4 POSITION REFERENCE SENSORS

Modern position reference sensors include the following:

a. GPS/DGPS/DARPS.
b. Artemis.

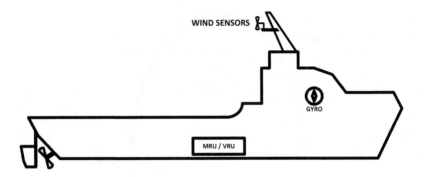

FIGURE 3.3   Environmental sensors.

    c. Fanbeam.
    d. Cyscan.
    e. HPR/HiPap.
    f. Taut wire system.
    g. RADius.
    h. RADascan.

3.1.5   Computer (RTC/DPC/Controller)

The computer here refers only to the DP computer/real time computer (RTC) or the controller. This is the central nervous system of the DP system, which upon taking all the inputs into consideration and based upon the mathematical model of the vessel, calculates the output which is known as the thruster allocation logic (TAL) (Figures 3.4 and 3.5).

3.1.6   MMI/HMI

The hardware which facilitates all the input to be entered, the displays and the dedicated computer for it, are all referred to as the man-machine interface (MMI), more often referred to as the human machine interface (HMI).

The HMI plays an important role for the efficient and safe operation of the system by providing the operator access to feed inputs for making optimum operational decisions. Most DP systems have arranged the devices in a way to assist logical operation and presentation of the required information. The hardware and presentations/displays have been designed to be user friendly. Due care has been taken to avoid accidental operation. This may be achieved by providing "double click" buttons and a combination of buttons, for example. Many DP systems already use such arrangements to prevent accidental operations.

For easy access, the buttons used are very often arranged in a way so that the operator can identify them quickly, and hence they are arranged in a group. For example, all the thruster-enabled buttons are located together.

FIGURE 3.4 DP controller (courtesy of Kongsberg).

3.1.7 OPERATOR

The operator is the seventh component of the DP system. A DP operator is generally trained to use the other six components so they can safely manoeuvre the vessels or ensure position keeping. The training of DP operators is managed by the Nautical Institute, London. However, there has been demand from a certain section of the industry for an alternate scheme of training envisaged by the Det Norske Veritas (DNV) which is generating lots of interest and debate.

During normal operations the bridge and the DP desk is manned by a DP Operator (DPO) and a senior DP Operator (SDPO). The combinations may also be, on some vessels, a DPO and a junior DPO.

Safe operations require that all DPOs on board are fully familiar and trained to safely undertake the operational activities. Every DPO is expected to be aware of the vessel's capability, the failure mode effects analysis (FMEA)/failure mode effects and criticality analysis (FMECA) and the other the annual trials report of the vessel.

FIGURE 3.5   DP controller (courtesy of Converteam).

Additionally, the experience shows that a DPO who are familiar with the vessel's power system, thrusters and sensors would prove to be an asset on board.

It is expected that all companies and employees maintain and make use of standard checklists for DP operations.

Appropriate procedures should ensure that the requisite checklists are maintained as per the company's DP operations manuals. As the DP operators are the key persons who help and contribute to the coordinated work involving various departments like deck, engine room, diving controls, ROV controls, gangway, crane and helicopter operations on board, the role played by them is very central and important to safe operations. There are company and industry guidelines available contributing to safe operations which every DPO must be aware of and follow.

Every engineer on board must be able to provide the necessary help and assistance to the DP operators by ensuring that the remaining six components of the DP system (other than the DPO), work satisfactorily at all times.

4 Six Degrees of Freedom – Basics of Dynamic Positioning

4.1 SHIP MOVEMENT

A free-floating structure at sea is subjected to three planes namely, vertical, lateral and longitudinal axes. These three planes result in six freedoms or axes of movement. A dynamic positioning (DP) system is concerned with the horizontal plane movements, and all the movements on the horizontal plane must be controlled. Although not controlled, the other plane movements are measured using various sensors for the purposes explained below.

Out of the six movements three are rotational movements and the other three are translations. They are explained below.

4.1.1 ROLL

When the vessel is subject to environmental forces from the sides, a movement is experienced on the longitudinal axis. This may result in a rotating movement of the vessel along its longitudinal axis and this movement is known as the roll (Figure 4.1).

4.1.2 PITCH

When the forces acting on the vessel result in movement about the transversal axis, the vessel is said to be pitching. The rotating movement of the ship along its transverse axis is called pitch. It is important to note here that pitch is measured but not controlled. Pitch is measured by an environmental sensor called vertical reference unit (VRU)/vertical reference sensor (VRS) or motion reference unit (MRU). Both the roll and pitch movements are fed to the controller unit to compensate or offset the roll and pitch movements of the position reference sensors (Figure 4.2).

4.1.3 HEAVE

When the ship is riding the crest of two or three waves, it is lifted upwards and when the waves travel far away or the trough comes in the ship moves down heavily, these movements are called heave. Heave is measured by the MRU and is not controlled by the DP system.

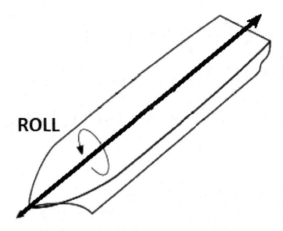

FIGURE 4.1 Roll movement.

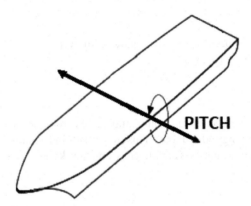

FIGURE 4.2 Pitch movement.

Although the roll, pitch and heave motions are not controlled, they are measured. The measurement of roll, pitch and heave are fed to the DP controller for the purpose of compensating the offsets of the position reference sensors. This is very important as without compensating the movements of the position reference system (PRS), the vessel movement will increase and the vessel may use thrust and power unnecessarily (Figure 4.3).

4.1.4 SURGE

Surge is the forward and backward motion of a vessel. Surge may be defined as a movement in the fore-and-aft direction, or fore-and-aft bodily movement of the vessel. This movement must be measured, monitored and controlled by the DP system using thrusters (Figure 4.4).

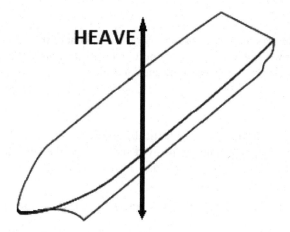

FIGURE 4.3 Heave movement.

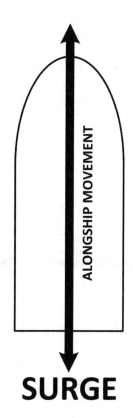

FIGURE 4.4 Surge movement.

4.1.5 SWAY

Sway is the sideways motion of a vessel. Sway movement of the ship is the athwart-ship movement when the ship moves to port and starboard bodily. The sway movement of the vessel needs to be measured and controlled by the DP system (Figure 4.5).

4.1.6 YAW

When the vessel is turning about its vertical/Z axis there is a change in the vessel's heading. This is then picked up by the gyro and given to the DP controller. This is generally referred to as yaw motion (Figure 4.6).

4.2 MOVEMENTS CONTROLLED AND MONITORED BY DP

The position and heading of the ship need to be controlled by the active thrust of the vessel. DP is concerned with the individual and automatic control of surge, sway and yaw. The ship's position is controlled by measuring the surge (X axis) and the sway (Y axis) movements of the vessel. These are measured by PRS and position measuring equipment (PME) (Figure 4.7).

Yaw movement of the vessel must be controlled. This is measured by the gyro. Gyro input is fed to the comparator unit and the controller then generates the set point for a certain thrust to be applied.

Measurement of surge and sway needs an accurate PRS (Figures 4.8 and 4.9).

For surge and sway control the position reference sensors measure the current position of the ship and is fed to the comparator unit. This is used to calculate the

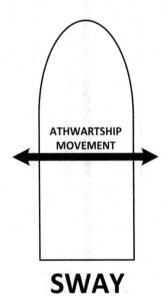

FIGURE 4.5 Sway movement.

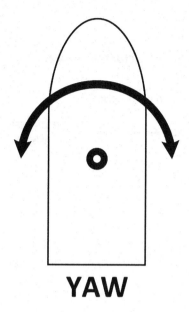

FIGURE 4.6 Yaw movement.

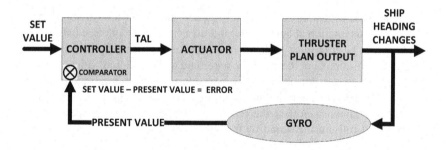

FIGURE 4.7 Yaw (heading) control loop.

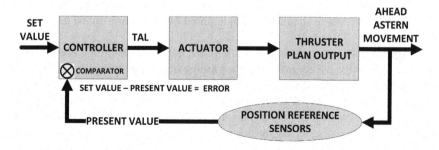

FIGURE 4.8 Surge control loop.

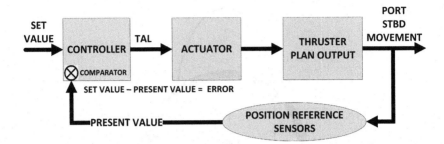

FIGURE 4.9 Sway control loop.

error. The error is analysed by the controller and the output is given. This is called the set point or thrust allocation logic (TAL).

Even if there is a complete failure of PRS, the DP may automatically control the heading of the vessel while the vessel position remains under manual control (joystick for surge and sway).

The other three degrees of movements are on the vertical plane. They are known as roll, pitch and heave:

- Roll: The roll of a ship may be defined as the movement around the vertical axis.
- Pitch: Pitch is described as vessel rotation about the athwartships axis, while roll is rotation about the fore-and-aft axis. Neither of these two movements can be controlled (although roll, may be dampened by active or passive stabilisation), but they must be measured and monitored with precision. This is necessary for accurate position referencing. Roll and pitch movements of the vessel are measured but not controlled. The measurement is used to compensate/offset the readings of the position reference sensors. Pitch and roll are measured by means of a vertical reference unit (VRS/VRU or MRU).
- Heave: Heave is described as vessel's bodily movement in the vertical or up-and-down axis/plane of the vessel. Heave is generally not used by the DP system, although the more sophisticated MRUs will be able to measure it. Once measured by the MRU, this may be monitored by other ship systems where, for instance, heave compensation is required for cranes or a diving bell.

Although the roll, pitch and heave motions are not controlled, they are measured and this is fed to the DP controller for the purpose of compensating the offsets of the PRSs, which is considered very important.

Most modern DP operator panels have a joystick (manual) and auto buttons giving the mode of DP control. In auto mode, all three (surge, sway and yaw) movements are controlled automatically. In joystick (manual) mode, control is affected manually by means of a joystick (surge and sway) and/or a rotatable control knob for manual control of yaw.

The operator may thus select any combination of manual and automatic control as desired:

- Auto yaw and sway, manual surge.
- Auto yaw and surge, manual sway.
- Auto surge, manual sway and yaw and many more similar combinations.

The present market has two types of DP systems as described below:

- PID controller-based DP systems.
- Model control-based DP systems.

DP systems based on the proportional, integral, derivative (PID) concept basically make use of a control loop feedback mechanism. This is referred to as a controller in the DP system. This is also a very commonly used control system for various industrial applications. An error value is calculated by the PID controller. This error is basically the difference between the set point (the desired value) and the currently measured value. This calculated error drives the control chain further and helps in achieving the position control of the vessel.

There are other DP systems which are based on mathematical model control.

As environmental forces act on the ship, various parameters are being changed and the model is being built.

A model-based system also helps the vessel to maintain her position and heading for a period of about fifteen minutes in case of total loss of PRSs. The majority of the systems use model-based control. In a model-based system, set points for position and heading are set by the operator and are then processed by the DP system. The computerised output, generally referred to as the TAL, will provide control signals. These output control signals will then control the thruster/propulsion output. The PRSs provide the input for surge and sway movements and the heading control input is provided by one or more gyro compasses.

4.3   MATHEMATICAL MODEL

Wind, waves and current, along with various operations-related forces work together to move the vessel away from the intended position. These movements are measured by the gyro compass, PRS, wind sensor, VRU or motion reference sensor.

The mathematical model is a tool to compute the difference between the set point and offset values. This difference is then used to compute TAL. which is then applied to the available thrusters and moves the vessel to the set point. The mathematical model contains the aerodynamic and hydrodynamic description of the vessel. This is used to calculate the vessel's response to the external disturbances (Figure 4.10).

The model or the mathematical model of a DP vessel may be defined as the mathematical description of how the ship may react or will move when environmental forces are acting upon it. The model is a hydrodynamic and aerodynamic description of the ship. This involves the vessel's characteristics such as shape, mass and drag. While developing

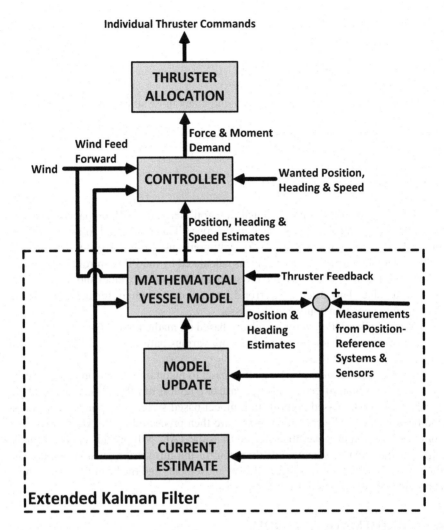

FIGURE 4.10 Mathematical model.

the mathematical model, care must be taken to ensure that the model is as accurate as possible so that the reaction to the forces acting on the vessel is also accurate.

The mathematical model, on a continuous basis, is affected by the forces that also affect the ship. Wind forces are directly measured as the function of wind force and wind direction. The effect of this is generally referred to as the force vector. To hold the vessel in position the DP system must be able to generate a thrust vector which must be equal and opposite to the force vector. The thrust vector is a function of TAL which includes revolutions per minute (RPM)/pitch of thrusters and propellers.

A number of parameters like the vessel mass, draft, hull form, propulsion type and arrangement, superstructure and location, etc., are utilised while preparing the mathematical model of the vessel.

While updating the model, a Kalman filter is generally applied. The method provides the effective noise filtering of the position and heading data. The algorithm utilises the measured data from reference systems and sensors along with the data predicted by the model. The calculated position and heading difference are now applied to the model update. The model also provides the so called "dead reckoning" for a short period when the input from the sensors has failed. If there is a severe system failure, such as the total loss of position or heading input to the system, the vessel model or the mathematical model will maintain thruster output settings.

4.4   DP SYSTEM

DP system may be defined as a system designed to automatically control and maintain a vessel's position and heading with a high degree of accuracy, within the difficult and hazardous surroundings of the marine environment, without the use of anchor or mooring lines, but by utilising the vessel's main propulsion and thrusters which is often referred to as the active thrust.

According to the International Maritime Organization (IMO), a DP system is defined as a system having three main components, namely the DP control system, the thruster system and the power system.

A DP system may set a target position, called a station, which can be a fixed or movable reference point on the sea floor, over which the vessel hovers to provide a stationary platform from which to carry out operations. A DP system also allows a vessel's easy and safe manoeuvring in the port or restricted area.

For any DP system to carry out these basic functions, the following subsystems and components would be required:

- Sensors for measuring external forces.
- A method of measuring the vessel's position.
- Sensors for measuring vessel heading.
- A system for calculating required counter forces.
- A system to generate the required forces.
- An arrangement to ensure man/machine interface.

A DP system measures deviations caused by environmental forces by the environmental sensors and the position reference sensors. This deviation is counteracted by forces generated by the thrusters. A DP system also aims at achieving minimal thruster activity and hence minimal use of power. A DP system is a combination of process wherein position control and heading control systems can maintain the vessel in position by controlling surge, sway and yaw.

A DP system's position control function makes use of the vessel's PME which is also referred to as the PRS. The DP system also uses inputs from the operator commands. The system then provides output as commands to the thrusters to maintain the position of the vessel at the desired location. Once the thrusters reach the commanded level, a feedback is sent to the controller. This is called a feedback control system or a closed loop system.

A DP heading control system uses the vessel's gyrocompass as the input to maintain the heading of the vessel in response to the external forces and operator commands.

4.5   ADVANTAGES AND DISADVANTAGES OF THE DYNAMIC POSITIONING SYSTEM

Some of the basic advantages of using a DP system are as follows:

- By using DP, the ship is fully self-propelled.
- Use of anchor or mooring lines is not necessary.
- Good manoeuvrability of the vessel is ensured, though she is maintaining position.
- When weather conditions change, the DP can respond faster.
- Operator command and operational requirements can be responded to quickly.
- Compared to other methods, water depth does not cause any restriction for operational capability and manoeuvrability of the vessel.
- Using a DP system can make operations more economical.
- Small tasks can be completed quickly, and the vessel can move to the next location/job.
- As a DP system does not make use of anchors/mooring lines, there is no danger of damage to the seabed flora and fauna.
- No risk of mooring lines crossing.
- Position and heading movements can be executed quickly.

Disadvantages of using DP vessels:

- Position-keeping capability lost if a component/subsystem fails.
- To operate and maintain a DP System, qualified personnel are required, thus incurring high day rates to hire qualified personnel.
- During operations, fuel consumption increases.
- The use of thrusters during operations may be risky for others like divers and remotely operated vehicles (ROVs).
- Strong tides, waters that are too shallow or too deep can pose a challenge to DP operations.
- A DP system still requires human intervention and is hence prone to human error.
- The more redundant equipment calls for more trained personnel, adding operational cost.
- Training the seafarers required to safely run DP operations pushes the cost up further.

4.6   WIND FEED FORWARD

Many environmental forces act on the vessel, but wind is the only force measured using wind sensors. The wind force should be measured accurately and fast enough so that the response from the thrusters can be directed to compensate the wind force. The faster the compensation takes place the lower the thruster force used, thereby reducing the use of power.

This should result in better position-keeping margins. This can be achieved faster by making use of a concept called "wind feed forward". This concept makes use of the fact that a DP system does not allow the vessel to move away from the required position, but counteracts the forces generated by the wind forces almost as fast as they are detected. There is a wind input directly given to the computer and the other goes through the mathematical model (Figure 4.11).

4.7   EXCURSION OR FEEDBACK OF SHIP POSITION

The DP controller gets inputs entered by the operator regarding the position and heading set points. These are calculated based upon the set points entered by the operator, the filtered position data, along with the differences between the speed needed and actual speed. This information is used to calculate the amount of thrust required.

This calculated output is then multiplied by the gain levels/settings to arrive at the thruster force needed. The same is represented by "thrust vectors" displayed on the appropriate screen/display. Similarly, the same kind of thrust output is used to slow down the movements of the ship if the it was moving at speeds faster than required.

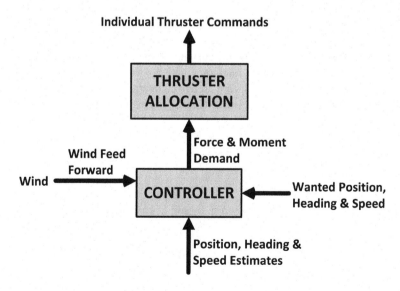

FIGURE 4.11   Wind feed forward.

The thrust force, thus calculated, may be influenced by the following:

- The amount of thrust required for restoring the position of the vessel. This is observed to be proportional to the difference between the required and the current heading and position.
- The difference between the required and the actual speed of position or heading change, which is also referred to as "damping thrust demand".
- To adjust the thrust demand, the gain factors are adjusted so as to optimise the performance of position-keeping capability and also minimising the power consumption. If the vessel is hunting around when on auto DP mode gain is one of the items to check and set to correct settings.
- If, during operations in auto DP mode, the vessel is hunting around, it is recommended to check and set the "gain" setting appropriately.

4.8 CURRENT FEEDBACK

The concept of current feedback is yet another important issue in dynamic positioning. It is a fact that the very basic function of the DP system is to control and maintain the intended position, heading, speed and rate of turn. Still it may happen that the position or heading, or both, go out of control and the vessel may lose position or heading (Figure 4.12).

It may be found that the resulting movement which caused the loss of position/heading may be due to some of the forces acting on the vessel, and these forces are not being measured. These unmeasured forces may be waves and the sea current, which have a sizeable impact on the position-keeping capability of the vessel which are not measured directly but rather calculated by the DP system and the result is displayed as the "DP current".

In a very simple way, this may be defined as the total sum of unknown or unmeasured forces affecting the vessel. The DP current is then used to calculate the TAL and that in turn would be used to generate command signals. The thruster output forces which are referred to as thrust vectors are used to counteract the unintended movements of the vessel. The DP current is determined/calculated by the mathematical model.

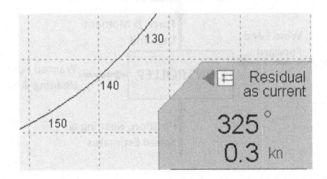

FIGURE 4.12   Current/DP current/residual as current.

4.9   THRUSTER ALLOCATION LOGIC

Once the controller receives the above-mentioned signals, it will calculate the output. This output signal (TAL) is used to allocate thrust to the available thrusters so that the known and the unknown forces can be countered. This is the process of controlling the controlled degrees of freedom, i.e., the surge, sway and yaw.

The TAL should result in the optimum use of the thruster force and hence, optimisation of the onboard power usage. This also controls extra wear and tear of the equipment and ship's machinery.

4.10   QUICK CURRENT OR FAST LEARN

Some DP systems call this phenomenon as the "quick current", whereas others refer to it as "fast learn". A quick current or fast learn facility gives a solution to the problems arising out of fast changes in the environmental conditions. The sea current is not measured but is calculated by a mathematical model. The updating of this mathematical model will decide how fast the response to changes will be made. If the environmental forces change rapidly, the mathematical model may not cope and there may be a delay in response. This may result in deterioration in position-keeping capability.

Most modern DP systems these days use a facility to update the mathematical model at a higher speed for a quick response to the environmental forces. It should be kept in mind that although this facility helps greatly in quick response, it is not one hundred percent perfect and may still lead to difficulties during quick weather changes.

4.11   KALMAN FILTER

The algorithm used to control the vessel's three degrees of freedom on the horizontal plane (surge, sway and yaw), uses vessel parameters. In this process, it is also important to know the parameters of the current, which otherwise are not easy to measure. The current speed and the direction relative to the ship are important.

However, the fact that all parameters may not be possible to measure correctly, require these to be estimated. This estimation is computed by the mathematical model. Disturbances like waves, which are not measured, also cause movements of the vessel and are required to be computed. Wave disturbance is taken into account by wave filters. Though computed very well, the mathematical model is not perfect. To increase the accuracy, the measured data and data obtained from the mathematical model are used together. These are further processed by the Kalman filter algorithm.

The Kalman filter makes use of a vessel's motion mathematical model for calculating measurements. The position data is received from various PRSs. This may be treated with the filters and they are accorded weightage according to their noise levels. The outcome of this mathematical model is used to update the vessel's position (Figure 4.13).

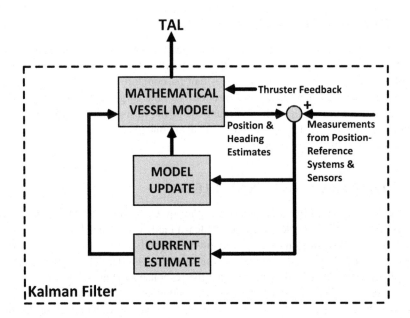

FIGURE 4.13   Kalman filter.

So, it may be observed that the Kalman filter helps in the estimation of vessel motion parameters and parameters of current. In case the measurements are not available for some reason the algorithm calculates a fairly accurate prediction of vessel motion parameters for some period of time and helps the vessel to remain in position.

4.11.1   WAVE FILTER

The wave filter can easily be defined as an extension of the Kalman filter. Wave filters can filter out high frequency waves.

The following three filters are commonly used:

* For disturbance on the longitudinal axis.
* For disturbance on the transversal axis.
* For disturbance on the vertical axis.

4.11.2   ADVANTAGES OF THE MATHEMATICAL MODEL

* The mathematical model is used to integrate multiple position measuring equipment (PMEs) and this results in refinement of the position keeping of the vessel.
* The Kalman filter uses data from sensors and other sources of the vessel details. This is used in the modelling process. The rudder is generally not included in the process of turning movement of the vessel.

- The filtering process is capable of identifying and rejecting data which do not seem to be good.
- The mathematical model is capable of maintaining the vessel's heading and position even after the input from sensors and PRS have deteriorated.
- Using the current data from the sensors and PRS, the mathematical model is capable of updating itself. This makes the positioning better.
- Filtered sensor signals help in noise reduction of sensors and thus the thruster activity. Data not meeting with the expected characteristics is identified and rejected.

4.11.3 MODEL CONTROL OR DEAD RECKONING (DR)

- There are occasions when the ship may lose position or heading inputs due to certain failures. The vessel may be still able to perform position-keeping tasks using automatic control by utilising estimated data. This estimated data is based on the conditions of the previous few minutes.
- Vessels have an extended operational window since positioning can be maintained during more severe weather conditions.

4.11.4 LIMITATIONS OF KALMAN FILTER

- At the design stage of the vessel, the conditions and parameters may not represent the initial design and may be based on conditions that may not fully represent the current operating conditions of the vessel.
- The tidal changes and sea current are usually not represented in the mathematic model.

4.12 DP OPERATIONAL MODES

There are a number of ways a DP-enabled ship may use to control the three degrees of freedom, i.e., the surge, the sway and the yaw. How the set points for speed and position are generated will decide the various modes the DP-enabled vessel may operate in.

The most modern DP systems provide access to the operator for these set points to be generated either by using a joystick lever or by using other dedicated devices generally referred to as human machine interface (HMI), or previously known as man machine interface (MMI).

The commonly utilised modes of the modern DP vessels are described below.

When the operator utilises the joystick lever to execute any movement in surge, sway or yaw, the mode is called "joystick mode". Some DP manufacturers in the past called this "manual mode". This must be clearly understood and must not be confused with the manual control of thrusters by dedicated thruster control levers. Usually the dedicated thruster levers are used in situations when there is a failure in HMI and/or the joystick lever system and the same cannot be used to generate the speed or position set points. It is worth noting here that when the thrusters are controlled by the individual levers this is not a mode of DP control.

- In "auto DP mode" position and heading control is carried out automatically by the DP system.
- In "mixed mode" the position may be controlled by using joystick and automatic options combined, as may be convenient in a given situation. Out of three movements to be controlled, even if one is being controlled by the joystick, many DP systems still prefer to call this mode "joystick mode" or "mixed mode".
- "Auto area position mode" is designed to keep the vessel within an allowed area. This mode also keeps the vessel within allowed heading limits and aims at minimising use of power automatically.
- "Auto track modes" are designed to keep the vessel on a specified track made of a set of way points. The mode may have sub-modes like low speed and high speed.
- Many DP systems have "autopilot mode". This mode is similar the autopilot function of a conventional ship. In DP vessels this mode permits it to follow a predefined path automatically.
- "Follow target mode" or "follow ship mode" is designed to help the vessel to automatically follow a target. The target could be a surface target or a subsea target.
- There is a need for many special vessels to have special modes for specific jobs and operations. This may be in addition to the modes mentioned above. A shuttle tanker and FPSO duo, for example, may have modes like "approach mode" and "Weathervane Mode" which are used for special purposes only.

The above modes are very commonly found in Kongsberg DP systems, but very similar modes may also be found in other DP systems as well. The modes used in Alstom/Converteam/GE Converteam DP systems are given below:

- Alstom/Converteam DP may have modes like joystick manual heading (JSMH). A joystick is used to control vessel movement fore/aft/port/starboard. For heading control the heading control knob is used. New systems may have a very similar joystick to that of Kongsberg.
- In joystick auto heading (JSAH) mode, the vessel movement fore/aft/port/starboard are controlled by joystick deflection. Heading control is automatic controlled by the computer about its centre of rotation (COR) as selected by the operator. It is mandatory to have gyro input (heading and rate of turn) for the heading control loop to function.
- Alstom/Converteam DP systems also have a mode called "transit mode". This mode is utilised when the vessel is moving from location A to location B.
- "Minimum power mode" in the Alstom/Converteam DP system is very similar to weathervane mode in other DP systems. This mode is helpful in reducing the use of power by reducing the thrust requirement.
- In "DP mode", the vessel's heading and position are maintained automatically. For this mode the DP system requires input from at least one position reference sensor. The step-by-step procedure of putting the vessel in DP mode/auto position mode is given below. (Figure 4.14).

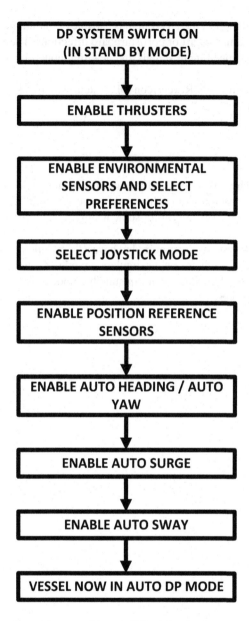

FIGURE 4.14 Steps for putting vessel in auto DP mode.

- "Auto track mode" uses a set of predetermined way points and fixed heading control. The vessel follows the track in low speed or high speed as selected by the operator.
- "Autopilot mode" in an Alstom/Converteam/GE DP system, provides ahead movement on pre-planned course and heading. Normally this mode is used as a transit mode, moving from one location to another.

- In "auto speed mode", the DP system is capable of maintaining zero or constant low speed in fore/aft or port/starboard directions. The provisions may be made to get speed input signals and heading inputs accordingly.

Take the example of a Rolls Royce Icon DP system which has the following basic modes of operation:

- Standby.
- Joystick.
- Positioning.
- Simulation: There is a need to have an offline mode which could provide simulated input/output data. This could be very helpful for onboard training. This mode is also helpful for some testing procedures.

In practice, most DP systems are required to have a "simulation mode". While in this mode, an operator can set the vessel to simulation mode and use it as a learning tool. It is important that the DP controls are in standby mode whenever it is intended to use the simulation mode.

5 Different Types of DP Vessels and Their Applications

5.1 TYPES AND APPLICATIONS OF DP-ENABLED SHIPS

There has been a significant increase in the number and application of dynamic positioning (DP) enabled vessels in the past two decades. The application of DP vessels has been increasing and new areas are being tried with DP applications. Historically, the DP applications have started drilling operations by keeling the vessel safely in position. The drilling operations may involve various activities such as coring, exploration drilling and production drilling (Figure 5.1).

Supply and standby vessels are used for replenishment of necessary logistics for the oil industry. Use of DP for such vessels and operations permit these vessels to work in close proximity to oil rigs, platforms, barges or other vessels with safety. These vessels can be classified into different types depending upon the tasks performed by them, such as:

- Anchor handling tugs.
- Anchor handling tug and supply vessels (AHTS).
- Offshore construction vessels (OCV).
- Remotely operated vehicle (ROV) support vessels.
- Diving support vessels (DSV).
- Standby vessels.
- Inspection, repair and maintenance (IRM or IMR) ships.

Some of these vessels may carry out combinations of these tasks and may be named multi-purpose support vessels (MPSV). This also helps the vessel to maintain her position and thus safe distance from and other vessels/structures/hazards.

5.1.1 ANCHOR HANDLING TUG SUPPLY

Towing, anchor handling, barge support, floating production storage and offloading (FPSO) support, moving and positioning drilling rigs transporting supplies and supporting drilling rig activities. These vessels also provide standby work and have firefighting capability (Figure 5.2).

Remotely operated vehicle (ROV) support vessels make use of DP systems for safely launching, retrieving and operating ROVs at various depths of water and

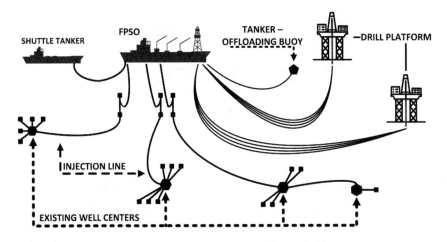

FIGURE 5.1   Offshore drilling.

FIGURE 5.2   Anchor handling tug and supply vessel.

doing different activities. The mode used more often is "follow sub" or "follow target mode" (Figure 5.3)

5.1.2   DIVING AND UNDERWATER SUPPORT VESSELS

Diving support vessels (DSV) make use of DP systems to safely position the vessel to ensure that the divers working below are safe. They mainly support diving operations by ensuring safe deployment and recovery of divers.

Divers may carry out different underwater interventions like inspection, installation, configuration, monitoring and recovery, etc. These activities may involve high risk.

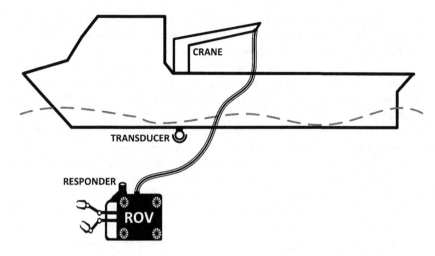

FIGURE 5.3 ROV support vessel.

During the job it is important that the vessel maintains, and when needed, should be able to move to an intended position safely (Figure 5.4).

There are hazards associated with dive operations and hence it is important that the identified hazards are mitigated by providing safety arrangements. One such arrangement in a DSV is the restriction of the diver's umbilical cord particular depths so that the diver must not get close to thruster beyond prescribed limits.

The length of the diver's umbilical cord is restricted to prevent him/her from being sucked into a rotating propeller. A standby diver is deployed to assist the working diver to handle his/her umbilical cord or to assist him/her in an emergency situation. Most of the DSVs prefer to use duplicated or triplicated DP equipment thereby ensuring that divers can be recovered in an emergency. When conducting diving

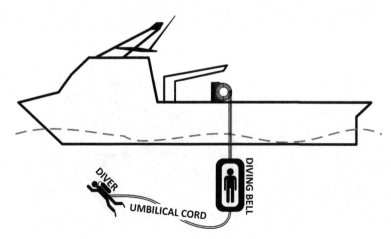

FIGURE 5.4 Diving support vessel.

operations from such a vessel on DP, the length of the diver's umbilical cord must be no more than the distance to the nearest thruster from the umbilical tending-point, minus five meters. This will ensure the safety of divers.

Pipe and cable-laying vessels also use DP systems to ensure the cable/pipe-laying jobs are carried out safely in any water depth and with changing weather conditions.

Multi support vessel (MSVs) or multi-purpose support vessel (MPSVs) are the workhorses of the modern offshore DP fleet, as they can be utilised for multitasking. Generally, the MSVs are DP Class 2 and can carry out the tasks assigned safely. The redundancy provided makes an MSV or MPSV a special vessel.

Accommodation vessels or "flotels", as they are commonly called, are the vessels providing accommodation services for the platform personnel (Figure 5.5).

Accommodation barges or flotels are usually together in place for months and hence have to have high level of redundancy to ensure position keeping. The general practice is that when the accommodation barge is directly connected to an oil producing rig with a telescopic gangway, it is expected to be a DP Class 3, but if placed outside the 500 metre zone the vessel may be a DP Class 2.

5.1.3 HEAVY LIFT VESSEL

Heavy lift or crane vessels require accurate positioning to place the equipment/ structure on the right position. Using DP, it is possible to carry out lifting operations safely. These vessels play an important role in subsea installation and accurate lifting operations. Heavy lift vessels are used to carry out offshore construction work by assisting in lifting heavy machinery and structures/equipment. Some of

FIGURE 5.5   Platform supply vessel.

the heavy lift vessels may be used to lift other vessels and for transporting them as may be needed.

Survey vessels using DP systems provide services for carrying out hydrographic or seismic survey activities. These vessels may carry long towing lines called "streamers" behind them and positioning or moving at a predefined speed is a challenging task, which may not be handled so well without DP.

Dredgers can perform better by utilising DP and increase operational safety. Dredging in defined areas and predefined tracks can be performed efficiently by using a DP-enabled dredger. The difficulties faced by normal vessels such as the drag force of pipes, water depth, etc., can be overcome.

Rock dumping vessels make use of a fall pipe which is connected to an ROV. The ROV positions itself making use of high precision acoustic positioning (HiPAP)/ hydro acoustic position reference sensor (HPR) system and thus helps the vessel to cover the pipeline or a structure with rocks. This saves and strengthens the pipelines/ structure from damage.

A FPSO vessel utilises weathervane mode to be in position so that the shuttle tanker behind her can off take the product safely. Both the vessels make use of dynamic positioning. "Weathervane mode" is used to maintain heading into the weather. Sometimes these vessels may use "anchor mode" to increase anchor effectiveness.

The main objective for a shuttle tanker to use DP system is to ensure safe loading, no unwanted disconnection and to maintain relative positioning to the FPSO aimed at avoiding undue strain/damage to the hose.

The DP system provides a distinct advantage so that an emergency disconnect can be avoided, the tanker can make a safe approach and maintain relative positioning. More often for these operations the vessel makes use of weathervane mode.

5.1.4 CRUISE LINERS

The latest types of vessels using dynamic positioning are cruise liners as they are not permitted to anchor to avoid damage to the flora and fauna under water. It is the DP system which helps them to be in a desired position so that their guests can enjoy the beauty of the place. A suitable DP mode may be used according to different situations.

5.1.5 OTHER SPECIALISED VESSELS

In addition to the above-mentioned vessels, dynamic positioning also helps more specialised functions like firefighting, military operations and search and rescue operations.

5.1.6 DP ADVANTAGES

By using a DP system, it may be possible to control three movements of the vessels, namely, surge, sway and yaw (position and heading), independent of external thrust.

The thrust used is captive thrust, produced by the thrusters, which is calculated by the thrust allocation logic.

This contributes to safer operations and the DP-enabled vessels are capable of working with more confidence in any work site.

A DP system may easily be used to change the position and/or heading of the vessel, hence allowing closer approach to the worksite. The vessel may eventually be placed closer to work sites and with optimum distance maintained from the thrusters, so that the working thruster/s do not hinder the operations in any way. The disadvantages include the continuous running of thrusters which may harm the divers, ROV or their umbilical cords.

6 DP Class/ Equipment Class

6.1 INTRODUCTION TO DP CLASS

This chapter discusses the various DP classes available on the market based on IMO/MSC 645. The major classification societies' notations are compared with the International Maritime Organization (IMO) and Norwegian Maritime Directorate (NMD) consequence class. A brief about DP system configuration is given on how systems are named as "Basic" or "Integrated" systems. The chapter also explains about the various makes of DP system available in the market.

It is recommended that the new IMO/MSC 1580 guidelines dated 16 June 2017 should be applicable to DP-enabled vessels constructed on or after 9 June 2017. As far as the old guidelines are concerned, they will continue to be applicable to ships and other offshore units that were constructed on or after 1 July 1994 but before 9 June 2017. It is important to note here that Section 4 of the new guidelines dealing with operational requirements will be applicable to all the vessels as appropriate.

It is the worst-case failure and the worst-case failure design intents which define the equipment classes, also known as DP classes. The equipment class or the DP class may be defined as given below:

a) As applicable to DP Class 1 or equipment Class 1, whenever there is a single fault occurring in the DP system, then the vessel position or heading or both may suffer a loss.

b) As applicable to equipment Class 2 or DP Class 2, a single point of failure may not affect the vessel's position or heading or both. This is achieved by the introduction of the concept of redundancy.

   Some static components may be accepted to fail in this class as they will not immediately affect the position-keeping capabilities of the vessel once they fail. For example, ventilation and seawater pipes, etc. Cooling water ventilation pipes which directly cool the running machinery are not considered in this category.

   For a DP Class 2 vessel, there may be the following single points of failure criteria but not limited to: All the subsystem/components which are in a continuously ready state, for example thrusters, generators, switchboards, remote controlled valves, network devices and various other communication and interfacing devices, are considered to be active components.

c) A loss of heading (yaw) or position (surge and sway) or both is not expected to occur for a DP Class 3 when there may be a worst-case single point of failure in the system. A worst-case failure or single point of failure for DP Class 3 may include the following:

All items as listed above for DP Class 2. Additionally, what is considered as normally static, other than the active components defined above, are considered to be failing.

For DP Class 3, one set of redundant equipment is expected to be installed in a separate compartment, safe from fire or flooding (A/60). This is an "A class" of bulkhead capable of withstanding fire or flooding for a period of 60 minutes.

It is important to note here that for the DP Class 2 and 3 a single inadvertent act by a member of staff may also be considered as a single fault. Such cases are considered if these are likely to happen.

A flag state verification and acceptance document (FSVAD) is issued by the relevant administration to a ship once the ship has been found satisfactory after an audit for this purpose.

6.2   MAJOR CLASS DP NOTATIONS

Depending upon the redundancy of the equipment installed on the vessel, a DP system is classified as DP1, DP2 or DP3. Some vessels are classified as DP Class 0 "zero" by some classification societies. DP Class 0 is not recognised by the IMO (Figure 6.1).

6.3   DP VESSEL CLASS/IMO DP CLASS

DP classification is based on guidelines from the apex body for maritime, popularly known as the International Maritime Organization or IMO. It is the Maritime Safety Committee (MSC) wing of IMO which is responsible for the regulations as published in the IMO/MSC letter 645 dated 6 June 1994 and the new IMO/MSC letter 1580 dated 16 June 2017.

6.3.1   DP Class 1

DP equipment Class 1 arrangements may not have redundancy. Once there is loss of certain component/s, this may result in a single point of failure. DP Class 1 ships will be designed to have automatic heading and position control of the ship.

IMO Equipment Class	NMD Consequence Class	DNV Class	Lloyds	ABS Class	BV Class	IRS Class
Not Recognised	Class 0	DNV-T	DP(CM)	DP-0	DYNAPOS SAM	NA
Class 1	Class 1	DNV-AUTS	DP(AM)	DP-1	DYNAPOS AM/AT	DP(1)
Class 2	Class 2	DNV-AUTR	DP(AA)	DP-2	DYNAPOS AM/AT R	DP(2)
Class 3	Class 3	DNV-AUTRO	DP(AAA)	DP-3	DYNAPOS AM/AT RS	DP(3)

FIGURE 6.1   Major DP classes.

A standard DP Class 1 system may consist of a minimum of one operator station fitted in the bridge console; the required number of environmental sensors, at least two reference sensors and one DP controller suitably located with required input-output devices. Both position references must not be of the same type to avoid common mode of failure (Figure 6.2).

The DP controller then calculates the output to control the installed thrusters, which help the vessel to main position and heading. The system may incorporate field devices or auxiliary field devices depending upon the configuration of the system. Some systems may have two operator stations, a controller and required sensors.

A DP 1 system may be configured as "basic" or "integrated/networked". This is explained in the following pages.

6.3.2 DP CLASS 2

DP equipment Class 2 has dual redundancy, and sometimes referred to as a dual redundant system. The dual redundancy has an advantage over single redundancy in that a single worst-case failure may not cause a loss of position or heading, or both. If there is a loss of one or more of the active components like thrusters, related remote-controlled valves, generators, switchboards, etc., may not cause a failure and loss of position or heading (Figure 6.3).

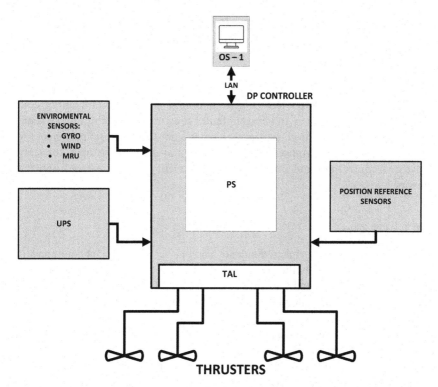

FIGURE 6.2   DP 1 system.

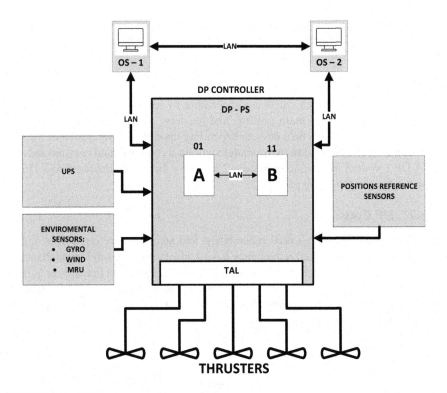

FIGURE 6.3 DP 2 system.

A DP 2 system is expected to have automatic position and heading control with complete redundancy in environmental sensors, position reference sensors, power and thrusters. All classes prescribe at least three reference sensors for a Class 2 vessel. Some high-end DP 2 vessels may have three operator stations and three controllers and thereby have the facility for voting and pooling of sensors.

6.3.3   DP CLASS 3

A DP Class 3 must have triple redundancy and one set of redundant equipment must be installed in the A/60 bulkhead area. An A/60 area can withstand fire or flooding for at least 60 minutes and thus in such an eventuality, the system may be utilised to move the vessel to a safe place, ensuring the safety of the vessel and of others in vicinity (Figure 6.4).

DP Class 3 systems are designed to have automatic and manual control of position and heading movements. The DP Class 3 vessel is expected to continue to hold its heading and position during and after a single worst-case failure is detected. The single point of failure may include fire and flooding, and hence one of sets of the redundant equipment is installed in A/60 bulkhead.

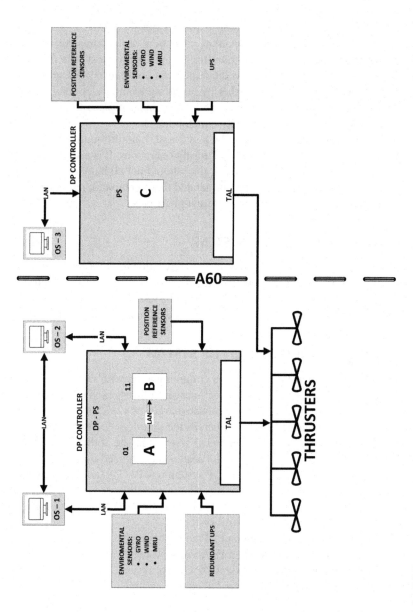

FIGURE 6.4 DP 3 system.

6.3.4   DP Class 0

Some classes also recognise a DP Class 0. The DP Class 0 has automatic heading control and allows position control of the vessel using a joystick. It is important here to understand that the DP class is assigned based on the redundancy.

As per the International Marine Contractors Association (IMCA), redundancy may be defined as the ability of a component, subsystem or the complete system, or system to maintain or restore its function after a single point worst-case failure has taken place. There are different ways redundancy can be achieved. For an example it is possible to achieve redundancy either by installing a number of components to suit the class requirements, or by having an arrangement in place to perform that particular function or task by an alternative method.

It is highly recommended that for DP Class 2 and 3, a minimum of three position references systems (PRS) is to be used during operations. It is also important to note that when two or more PRSs are used, they should not all be of the same type, working on the same principle. It is recommended that to maintain redundancy that at least two PRS should be working on different principles.

6.4   BASIC/HARDWIRED DP SYSTEM

A stand-alone or basic DP system makes use of conventional signal cables and serial lines for interface with other systems like the power plant and thruster controls. A basic DP system may be marked as 11, 21 or 31; the last digit "1" indicates a basic system, whereas the first digit 1, 2 or 3 indicates the number of DP computers.

6.5   INTEGRATED/NETWORKED DP SYSTEM

A networked or integrated DP system has communication with other systems like marine automation in the engine room machinery and thrusters via dual redundant ethernet/LAN. An integrated DP system may be marked as 12, 22 or 32; the last digit "2" indicates an integrated DP system, whereas the first digit 1, 2 or 3 indicates the number of DP computers. So, effectively a DP system may be of different combinations and specifications depending upon the design and intended use.

The list below explains the Kongsberg DP systems (K-Pos DP family) for example:

- DP-11: One DP control system and stand-alone single DP control system.
- DP-12: Integrated single DP control system.
- DP-21: Stand-alone dual-redundant DP control system.
- DP-22: Integrated dual-redundant DP control system.
- DP-31: Stand-alone triple-redundant DP control system.
- DP-32: Integrated triple-redundant DP control system.

7 Basics of Electrical Propulsion and Thruster Types

7.1 BASICS OF ELECTRICAL PROPULSION AND THRUSTER TYPES

Most modern dynamic positioning (DP) vessels are fitted with diesel-electric propulsion. The ship's diesel engines drive the generators supplying power to the electric motors for the thrusters.

Many vessels also continue to deploy a mixed propulsion system consisting of the main propellers and the required number of thrusters. The main propellers are fitted in the aft of the vessel. This arrangement may be assisted by and one or more thrusters which are used to manoeuvre the vessel. Thrusters may be designed to have fixed direction, as the case of tunnel thrusters which are mounted in a tunnel running through the hull. Additionally, there also may be azimuthing thrusters, which are mounted below the vessel and can be rotated 360 degrees. Generally, these propulsion/thruster devices are electrically powered.

7.2 DIESEL ELECTRIC PROPULSION FOR DP VESSELS

A modern DP-enabled vessel may have propulsion and thruster controls consisting of the following:

- Manual thruster control using individual levers, with arrangements to have separate control of each thrusters installed onboard.
- Automatic course control by using the available modes such as, standby, joystick, auto position, auto sail or auto pilot. The suitable mode may be utilised depending on operational status and type of job assigned.
- While maintaining position in field, the DP vessel may use a mode of DP system to automatically hold position by controlling the thrusters and propulsion system.
- Some vessels may use a combination control to assist the mooring operation using thrusters (Figure 7.1).

On any vessel thruster and propulsion control are crucial to safe operation as they play a vital role in ensuring the amount of thrust is available in the right direction and at the right time. When the vessel is in auto position mode or DP mode this decision

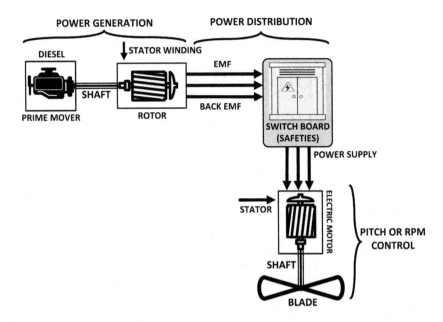

FIGURE 7.1 Diesel electrical propulsion system.

to generate the thruster allocation logic (TAL) is taken by the DP controller. While in other modes the intervention of the DP operator is crucial.

Designing a reliable and appropriate thruster control system is obviously important but the onus still lies with ship's crew to understand the control functionalities of these machines, along with power management and other related equipment. Propulsion/thruster control may employ either speed or pitch control.

The details below given are more suitable for power generation and distribution for a diving support vessel (DSV). Diesel-electric propulsion for a generic DP vessel, electrical power may be generated as given below.

7.2.1 Power Generation and Distribution

Normally, diesel generators may be installed in two separate locations and sometimes in separate engine rooms. Diesel generators generate power and send it to a HV switchboard. Generally, the switchboard, also is referred to as "bus", is connected to another switchboard by a bus switch or a bus tie breaker. The bus bars inside a switchboard are arranged according to a design based upon the power generation and distribution requirements (Figure 7.2).

The bus tie breaker on a typical DP 2 DSV is kept open during operation, thus providing redundancy and helping to avoid a complete blackout. When the bus tie breaker is closed, it makes the two switchboards function as one. This can be done to make the best use of the power available and the power required for a particular situation (Figure 7.3).

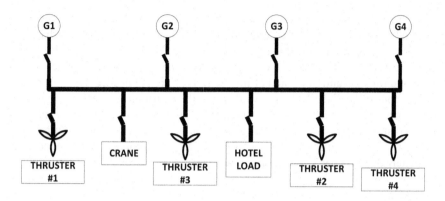

FIGURE 7.2 Non-redundant power generation and distribution.

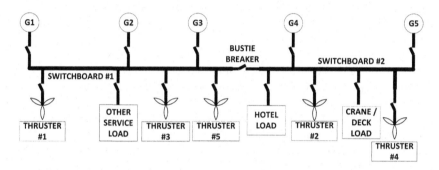

FIGURE 7.3 Redundant power generation and distribution.

Each busbar is designed to provide power to at least one main propeller/thruster, and at least one thruster may be a bow or stern thruster considering the redundancy concept. This arrangement provides safety in case of a power failure and bus malfunction.

Conventional diesel-electric propulsion of a DP-enabled vessel has a diesel engine running an electrical alternator/generator. Depending upon the size of the vessel, there may be a number of such arrangements to adequately power the vessel. Electrical power consumed by the thrusters is supplied by the power bank being fed by the diesel engines.

The switchboard is an essential part of a diesel-electric propulsion system. The alternators/generators feed the switchboard, thereafter the switchboard distributes the power to various consumers, the thrusters being the major and important consumers.

The switchboard or bus is also installed in separate spaces to meet the class requirements. These switchboards/buses, as they are commonly referred to, as are connected by a bus switch, which is commonly known as bus tie breaker. The bus switch is opened to isolate the two halves of the switchboard so each can operate independently of the other. When the bus switch is closed, the bus bars connect the two halves.

Each busbar provides power to one main propeller, and at least one thruster at the bow and stern. This provides redundancy should a fault develop in one busbar.

One of the advantages of this type of operation is the cost saving on fuel. Another advantage is the ability to take the generator online when the need arises, due to increased thrust requirement and offline when they are not needed.

The controller of the DP system calculates the thrust allocation logic indicating how much thrust is needed and the thrust needed to be generated by which all available thruster/s. This will be used to control the vessel's movement for surge, sway and yaw. TAL minimises thruster use, thereby, fuel consumption, wear and tear and ultimately overall operational costs are reduced.

Thus, thrusters play an important role in maintaining the vessel in position. Depending upon the design philosophy of the DP vessel, the thruster may be used for position keeping during operations as well for propulsion purposes while moving from one location to another.

Thrusters have to be reliable and efficient to control the vessel effectively. The appropriate thruster will have to be selected based upon the following:

* What the vessel is going to be used for? – the role of the vessel.
* How big is the vessel? – the size of the vessel.
* Where the vessel is going to operate? – the operating conditions.

Based upon the features of the vessel, expected operational conditions, type and size of the thrusters, a computer programme may be used to plot a graph to indicate the capability of the thrusters. This plot is known as the capability plot. Together, the thrusters of the vessel should be able to provide the required thrust and redundancy for the class of the vessel.

The thruster should be located in such a way that its operation should not interfere with the other thrusters or important sensors or diving system. If required, it should possible to stop/deselect the thrusters from the DP control panel. Any method of stopping the thruster must not result in an inadvertent operation.

DP vessels have a defined priority for the use of the thrusters. In case the total required thrust is not developed by the available thrusters to control all the axis movements, the DP system will have the following priorities:

* First priority will be heading control.
* Second priority will be position control.

7.3 THRUSTER RESPONSE AND ACCURACY

How well a DP vessel maintains her position and heading would ultimately depend upon the response of the online thrusters. If the thrusters are slow to respond or respond erroneously, the result may be undesired ship movements. Every class has its own standards, and the following details are indicative of acceptable performance levels:

* Response of pitch/speed control of thruster – zero to full thrust within 8 seconds (small thrusters).

- Zero to full thrust within 15 seconds (large main propeller).
- Thruster azimuth control – 0 to 180 degrees of azimuthing within 15 seconds (2 rpm).
- Pitch/RPM of thruster – ±2% (maximum pitch/thrust).
- Accuracy of Azimuth movement – ±1.5 degrees.

Even though thrusters are designed to meet certain specifications, they are still prone to failure. In case of a pitch, azimuth or speed control defect or malfunction, or there are errors in the control circuit, the thrusters may still continue to function either:

- The same way as it was functioning at the time of failure.
- In the event of a failure, the pitch or speed automatically goes to zero.
- Or the thruster automatically deselected from DP control or stopped.

Care is taken at the design stage so that in no case should a thruster go to maximum thrust condition (Figure 7.4).

7.4 THRUSTER MODES

Thrusters for marine use are designed to be controlled so that they work together in fixed positions or over restricted ranges as per requirement. When an appropriate mode is selected, fine control of DP vessel positioning is possible. This would also ensure that the thrusters are not running unnecessarily and thus saving power.

Not all thrusters may have the modes as described below but are applicable in particular to azimuth thrusters and propellers. Tunnel thrusters are not included as they need to be controlled at the port/starboard.

- Fixed mode.
- Biasing mode.
- Push/pull mode.

When a thruster is set to operate over its designed range (full range), i.e. 360° for an azimuth thruster, port or starboard for a tunnel thruster, it is usually referred to as "Free Run" or "Free mode".

7.4.1 FIXED MODE

In DP systems, the thrust allocation method is very commonly used. This process is based on the hybrid optimization process. The algorithm helps to calculate the thrust required for any given situation. This algorithm may consist of various programmes. While in fixed mode, a thruster is able to give continuous thrust (speed or pitch) in prevalent weather conditions. When an azimuth thruster or a group of azimuth thrusters are fixed to a predefined angle, these then provide thrust to a fixed angle. This arrangement is called the fixed mode of thrusters. This mode can significantly save the continuous changes in azimuth angles of the thruster.

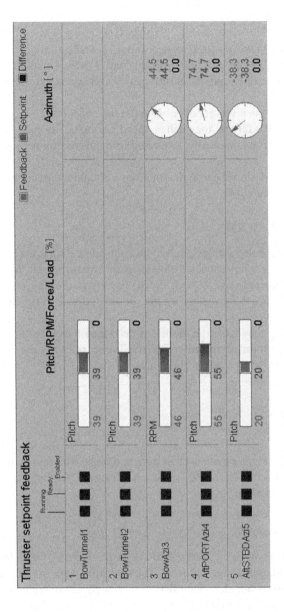

FIGURE 7.4 Thruster command and feedback (courtesy Kongsberg).

7.4.2   Thruster Biasing

When thrusters are biased, thrusters or a group of thrusters are placed in opposition to each other. Generally, this mode is applied to a pair of azimuth thrusters. When the thrusters are hunting during the time the vessel is experiencing low weather, there is a tendency that that TAL may not be able to calculate the correct amount of thrust and its direction. This may result in the continuous rotation of the azimuth thrusters. To take care of such a situation, the pair of thrusters may be selected in biasing mode (Figure 7.5).

7.4.3   Push/Pull Modes

Some propellers and rudders may have this mode of operation.

By using this combination, the resultant thrust may be provided in the sideways direction. Propellers and rudders can provide sideways thrust by using push/pull mode. This mode may also be referred to as "port ahead" and "starboard ahead". In push/pull mode, one propeller runs ahead, and the other is allowed to run astern. Generally, the rudder is allowed to operate at full rudder angles in light weather and at restricted angles in bad weather (Figure 7.6).

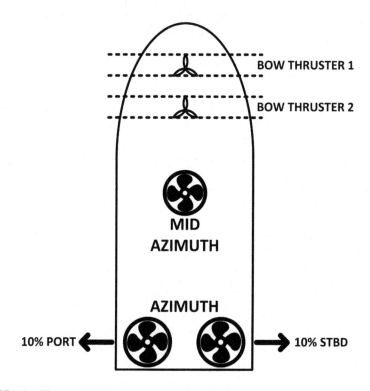

FIGURE 7.5   Thruster biasing.

FIGURE 7.6   Thruster push pull mode.

7.5   DIFFERENT TYPES OF PROPULSION UNITS AND THRUSTERS

With the demand for technologically advanced thrusters increasing, there are a plenty of makes available on the market. The commonly used thrusters in DP are as follows:

- Main propellers.
- Rudders.
- Tunnel thrusters (bow thrusters and stern thrusters).
- Azimuth thrusters.
- Podded thrusters (azipods).

Some specialised vessels may use special thrusters and they are also becoming popular. Some of them are:

- Gill jet thrusters.
- Voith Schneider propellers/thrusters (VSP).
- Shaft propellers.

In diesel-electrical propulsion, a variable speed electrical motor normally drives a shaft propeller. If a horizontal motor is used, then an arrangement is made to connect this directly to the shaft that is driving a propeller. This conventional arrangement is considered to be an easy and robust system which can generate maximum bollard pull for the same power. This system has a disadvantage in that to install such systems there is a need to have enough space to accommodate shafting arrangements (Figure 7.7).

If there is not enough space, horizontal installation may not be suitable and hence there may be a gear arrangement made which makes the system compact with high speed. Bigger vessels may usually have shaft propellers and use high lift rudders with this arrangement, otherwise they may be designed with a gear coupling which allows for increased rotational speed of the motor and results in a more compact unit. Shaft propellers are suitable for bigger vessel applications such as shuttle tankers. Such vessels may use high lift rudders to achieve sideways thrust.

7.5.1 Tunnel Thrusters

The thrusters fitted to the ends of the vessels, and usually inside a tunnel, are known as tunnel thrusters. They may be also called bow tunnel or stern tunnel thrusters according to their location.

Tunnel thrusters are designed to produce thrust in a transverse direction which may be used to either turn the vessel around its rotation point or may be used to achieve athwartship movement of the ship (Figure 7.8).

The tunnel thrusters are usually driven by an electrical motor, which may be fitted with a L shape gear arrangement or a Z shape gear arrangement. The thruster may be designed to be either pitch control or speed control. Some very specialised vessels may be designed to have both types of thrust control.

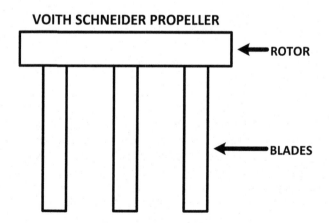

FIGURE 7.7 Voith Schneider propellers/thrusters (VSP).

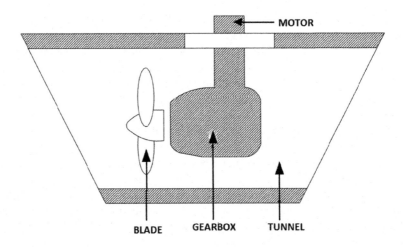

MOTOR

BLADE   GEARBOX   TUNNEL

FIGURE 7.8 Tunnel thruster.

7.5.2 AZIMUTH THRUSTERS

Azimuth thrusters are special type of thruster which have a rotatable device for production of thrust in any direction. The main drive motor for these types of thrusters may be mounted vertically and drive a shaft gear arrangement suitably designed to meet a vessel's thrust requirements. Sometimes an L shaped or Z shaped installation with a suitable RPM control system, such as variable frequency drives, help in reducing losses.

The main advantages of azimuth thrusters are easy manoeuvrability, better efficiency and effective utilisation of ship space. Azimuth thrusters also known to have low maintenance cost (Figure 7.9).

7.5.3 AZIPODS

An azipod is a special type of thruster from ABB. It is a podded propulsion system, azimuthing/rotating through 360°. Azipods may use a single- or double wound AC motor. Usually such motors will have a very small shaft, thereby reducing losses and improving efficiency of thrusters.

The drive motor drives a fixed-pitch propeller. To control the thrust, the RPM of the azipod motor is controlled by deploying a frequency converter. Industry considers that azipods can be used for different kinds of vessel applications (Figure 7.10).

As azipods have been proven to be reliable efficient, and space-saving, these are suitable for different types of vessels. The challenge to supply electrical power to a continuously rotating motor is taken care of by using slip rings for 360° operation. As there is no pitch control requirement and no gear transmission this solution for 360°propulsion is proving to be an ideal solution for different types of vessels.

Comparison of Diesel-Mechanical (DM), Diesel-Electric (DE) and Hybrid

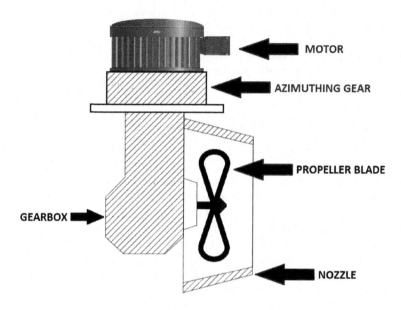

FIGURE 7.9   Azimuth thruster.

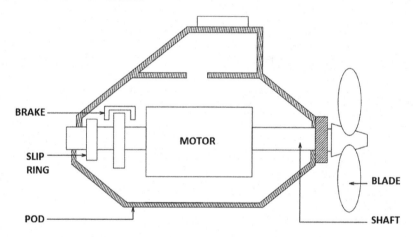

FIGURE 7.10   Podded azimuth thruster.

7.5.4   DIESEL-MECHANICAL (DM) PROS AND CONS

Pros:

- High efficiency at maximum load.
- Most bollard pull for the money.
- Rigid and durable workhorse design.
- Potentially longer mean time between failure (MTBF) than for the DE system.

Cons:

- Massive zero-pitch loss when idling.
- More fuel consumption and pollution at idle.
- Less power station redundancy.

7.5.5   DIESEL-ELECTRIC (DE) PROS AND CONS

Pros:

- High efficiency at idle and low to medium load.
- Multi-engine DE configurations have improved power station redundancy.
- Easy to apply the proper amount of power to all relevant situations.
- Flexible and automated operation.
- Improved manoeuvrability if azimuth-type main propulsion is used.
- Easier to design optimal arrangements below the main deck.
- If there is a straight shaft line propulsion system instead of an azimuth system, then there is potentially more bollard pull at the same power.

Cons:

- Lower efficiency at maximum load.
- Higher initial cost than DM.
- Potentially shorter MTBF than DM.

7.5.6   HYBRID DM/DE PROS AND CONS

Pros:

- This arrangement uses the best of DE and DM in one system making it more efficient.
- Lower initial cost than DE for large installations.
- Good redundancy.
- High efficiency at idle and low to medium load.
- Improved efficiency over DE at high loads.
- Easy to apply required amount of power to all relevant situations.

Cons:

- More complex installation than DE and DM.
- More complex user interface.
- Higher initial cost than DM.

It is always a good idea to have a medium-sized DM and DE systems in one large hybrid propulsion. This may be a suitable solution for a modern DP vessel and result in a better solution.

All DP professionals should have a fair idea about the efficiency of the propulsion system they work on. A diesel-electric system is normally considered to have an efficiency between 0.88 and 0.92 at full load. It is important to note that the efficiency of a diesel-electric propulsion system is influenced by the system load.

7.5.7 PRECAUTIONS WHILE WORKING ON CPP DRIVES

The following precautions must be exercised while working on a controllable pitch propeller (CPP) drive; this will help to ensure safe operations:

- While operating from "remote" ensure that the pitch angle is checked for ahead, astern and stop position.
- While operating from "emergency position" check pitch angle and position indicator.
- The hydraulic system is a very important part of the system and hence care must be taken to ensure that the system does not have any leakages.
- Maintaining an adequate oil level in the tank is very important and is a must for safe operation. The alarms for this must be tested for good operation.
- If the level and leaks are taken care of, it will be easy to maintain the required pressure for the operations. Maintaining the required pressure is important to maintain the desired pitch angle.
- It is safe a practice to start the CPP thruster at zero pitch. Safety measures are incorporated to ensure this. (If not done, the starting load may be beyond normal limits and unsafe)
- Check all the parameters at regular intervals (as recommended by the original equipment manufacturer (OEM). Also check the bearing temperature regularly.
- Analysis of hydraulic oil used in the CPP system to be carried out onboard using test kits provided to check the condition and water ingression to keep it safe foreign material.
- To maintain the thruster in good health it is recommended to run it at constant speed even at reduced loads and the thrust to be controlled as required by changing the pitch. This helps to maintain the efficiency of the CPP thruster.
- All precautions to be taken that no pollution is created by oil leakages from thrusters and for this daily checks are to be carried out "out board". Any trace of oil may be an indication of a leakage. This must be attended to as a priority.
- A CPP is considered a more complex installation as compared to the fixed pitch propeller (FPP). This demands that all concerned engineers and technicians are appropriately trained.

8 Thruster Controls and Automation

8.1 INTRODUCTION

The position-keeping capability of DP-enabled vessels depends upon the capability of the installed propulsion system, including thrusters. Generally, all the thrust-generating devices on board are referred to as the thrusters, which include the main propellers and the rudders.

According to DNV (Rules for Ships), a thruster system of a DP vessel will include all components, subsystem/s and system/s which are required to supply the DP system with thrust, force and direction. Keeping the above in mind, a thruster system may consist of:

- A drive unit with the required auxiliary systems, which may include piping arrangements.
- As system for thruster control, including the cabling and cable routing.
- The propeller/s and rudder/s, if under the control of DP system, are also known as thrusters and should follow the above.

It may be observed that in the majority of the DP vessels are that of diesel-electric propulsion type, most of the drives installed are powered by AC electrical motors. The techniques utilised to control the thrust required, either a controllable pitch propeller (CPP) or speed control methods, may be utilised depending on the design of the vessel. In the case of fixed pitch (FP), AC variable frequency drives also known as variable speed drives (VSD) are utilised to control the RPM and the thrust direction.

While using variable frequency drive (VFD), there is no need to have bulky pitch control system. Depending on the purpose and the intended use of the vessel during her lifetime, the type of thruster used will be decided and installed.

8.2 THRUSTER RESPONSE AND ACCURACY

As discussed in the previous chapter each thruster installed must meet the designed response standards and accuracy. Please refer to the appropriate guidelines and records that may be available in the trial reports.

Even though thrusters are designed to meet certain specifications, they are still prone to failures. In case of a pitch, azimuth or speed control defect or malfunction, or there are errors in control circuit, the thrusters may still continue to function:

- The same way as it was functioning at the time of failure.
- In the event of a failure, the pitch or speed automatically goes to zero.
- The thruster automatically deselects from DP control or stops.

Care is taken at the design stage that in no case should thrusters go to the maximum thrust condition.

Some specialised vessels may use special thrusters and the same are also becoming popular. Some of them are as mentioned below:

- Gill jet thrusters.
- Voith Schneider propellers/thrusters (VSP).

Rolls-Royce propulsion products used in the offshore/marine market are available in the following brand names:

- Kamewa™.
- Kamewa Ulstein™.
- Ulstein Aquamaster™.
- Bird-Johnson™.

Some of the popular thruster models from Rolls-Royce are mentioned below:

- Ulstein Aquamaster™ – azimuthing retractable thrusters.
- Ulstein Aquamaster™ – azimuthing pulling propeller.
- Kamewa Ulstein™ – tunnel thrusters.
- Mermaid™ – azimuthing podded propulsor.
- Kamewa™ – waterjets (A series and S series).
- Ulstein Aquamaster™ – combined azimuth/side thrusters.
- Kamewa™ – axial flow waterjets, FF-series.
- Ulstein Aquamaster™ – azimuthing contra-rotating propellers.
- Ulstein Aquamaster™ – azimuthing underwater mountable thrusters.
- Ulstein Aquamaster™ – azimuthing swing-up thrusters.

Wärtsilä propulsion products used in the offshore/marine market are available in the following brand names:

- Wärtsilä EnergoFlow – propellers.
- Wärtsilä EnergoProFin – propellers.
- Built-UP – propellers.
- Wärtsilä – propellers for coastal and inland waterways.
- Wärtsilä – fixed pitch propellers.

- Wärtsilä – controllable pitch propeller systems.
- Wärtsilä – transverse thrusters.
- Wärtsilä – underwater mountable thrusters.
- Wärtsilä – retractable thrusters.
- Wärtsilä – steerable thrusters.
- Wärtsilä – modular and midsize waterjets.

Similarly, there are other well-known brands in the marine/offshore propulsion and thruster market. A few of them are given below

- ABB.
- Thrustmaster.
- Kawasaki Heavy Industries.
- Veth Propulsion.
- Tees White Gill Thrusters.
- Brunvoll.
- CAT (Caterpillar).

8.3 THRUSTER CONTROL AND AUTOMATION

A modern DP vessel is expected to have independent manual thruster controls.

This is to be achieved by having an individual lever for manual thruster operation.

By use of a separate lever for each thruster, the manual control of any thruster must include independent control of the thruster. The manual control should include functions like start/stop, pitch/rpm control and azimuth control where applicable. The manual mode is a closed loop (follow up) mode. In a closed-loop control system for thrusters, feedback signals and command signals are compared and the difference of is used to generate a command accordingly (Figure 8.1).

Sometimes exceptions by certain classification societies are provided, if a safe backup/changeover is provided. The requirement for manual thruster control is applicable to all types of thrusters and must always be tested before operation as per the checklist. As a good design practice, no single failure in a thruster control system may result in an increase in thrust output. The failure must not also generate a command to rotate the thruster in azimuth mode (Figure 8.2).

The thruster controls are designed such that in the event of thruster failure it should result in "Fail-Safe" situation. Fail-safe ensures that due to this failure, the vessel is not likely to lose heading or position, or both, as the fail-safe mode will maintain the output for some time and generate a warning /alarm accordingly (Figures 8.3 and 8.4).

8.3.1 Wärtsilä Thruster Control

Wärtsilä's thruster control system, commonly known as Platinum Thruster Control System consists of various autonomous thruster control units. These units are connected using a network system to the DP controls.

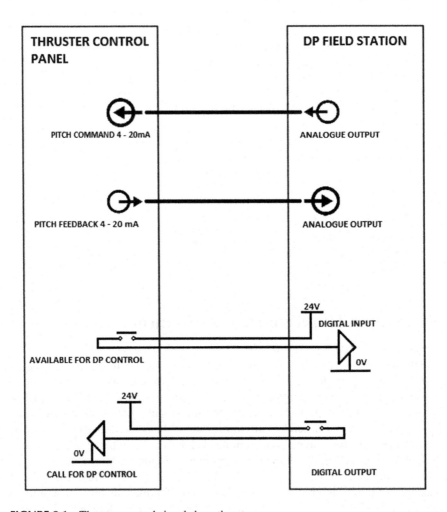

FIGURE 8.1   Thruster control signals bow thruster.

A control lever for each thruster is installed in the bridge. Thus, this lever can be used to control the speed and direction of thrust. Optionally, this may be duplicated in the control room using signal processor units (SPUs). This may be supported by fibre-optic cable or network cable which is independent of the DP network. DP input/output devices may be installed in the thruster control panels and DP controller.

This can be used to make the system "integrated" by using "Platinum DP" along with a full backup which is called an independent backup. Wärtsilä calls it an automatic thruster control system operator workstation (ATOW).

This integration helps in providing better thruster control and monitoring. This monitoring also includes alarm events and logging of data for the main equipment and the required auxiliaries.

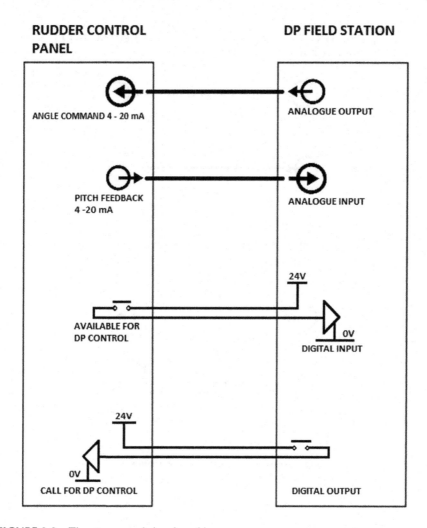

FIGURE 8.2 Thruster control signals rudder.

8.3.2 KONGSBERG THRUSTER CONTROL SYSTEM

A Kongsberg thruster control, more often referred to as K-Thrust, is a remote-control system to independently control all electrically driven propulsion devices. K-Thrust is designed to offer manual remote control of all thruster devices. Depending upon the vessel's requirement the K-Thrust can be configured to have following control modes:

- Lever/manual thruster control.
- Joystick control with an option to have automatic heading control.
- Auto piloting.
- Automatic position keeping (automatic control of surge, sway and yaw).

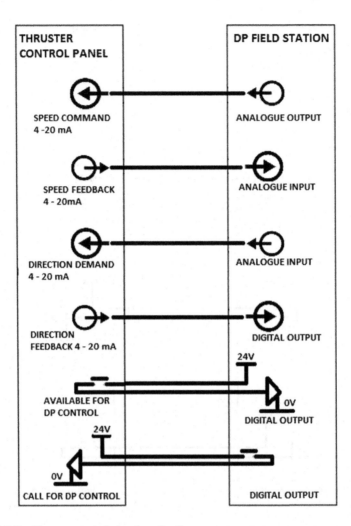

FIGURE 8.3 Thruster control signals azimuth.

Options may be provided to control the above from different locations including the bridge of the vessel.

8.3.3 SIEMENS THRUSTER CONTROL SYSTEMS

Like any other thruster control system, the Siemens thruster controls aim at active control of thruster drives. The thrust needed should be controlled through operator command or by the output of a DP controller usually referred to as thruster allocation logic (TAL).

The Siemens thruster control system is based upon an advanced control algorithm. All possible vessel characteristics and dynamics are calculated. This information is then used to compute the thrust necessary to counteract a given situation.

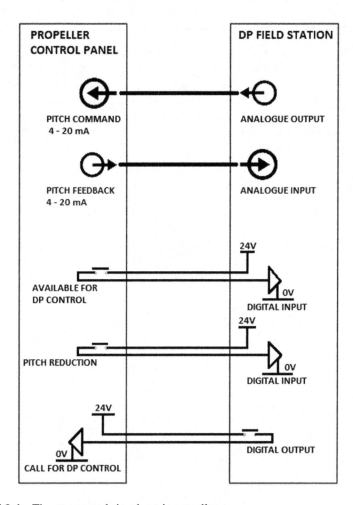

FIGURE 8.4 Thruster control signals main propeller.

PRS and other sensors are used find out any deviation from the intended position/ heading and accordingly the thrust requirement is calculated after updating the vessel control algorithm.

8.3.4 CAT (CATERPILLAR) MARINE

This microprocessor-based control system provides the operator with information and full control of all propellers, thrusters and engines installed onboard. The thruster control system provides:

- Redundancy and graphic displays for the operator.
- Control for the load of the main engines.
- Interface to DP systems.

8.3.5 EMERGENCY STOP FOR THRUSTERS

According to existing rules each thruster should have an emergency stop facility at
the main DP control station. This arrangement must be capable of shutting down
each thruster independently. The emergency stop arrangement must also work in
all modes of operation. The emergency stop facilities must use separate cables for
reach thruster. Also, an alarm must be activated once there is loop disconnection
detected. Thruster emergency stops must be tried out and records must be maintained (Figures 8.5 and 8.6).

8.4 THRUSTER MONITORING AND ALARM

All DP systems are designed to provide thruster monitoring. This monitoring is done
at the DP controller or the process station. The DP officer on watch is expected to
continuously monitor by looking at various indicators and alarms/warnings. This
can be an indicator of thruster health. Generally, the following parameters are monitored to see if the thrusters are working satisfactorily:

- Command and feedback of RPM, pitch, azimuth.
- CPP thruster's hydraulic oil level, pressure and temperature, etc.
- Motor temperature and thruster load.
- Semiconductor converter coolant leakage.
- Temperature of the converter controls.

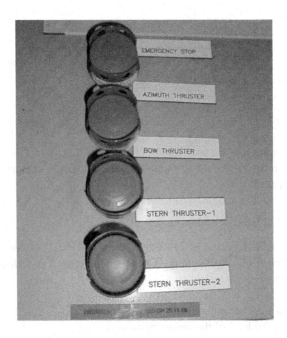

FIGURE 8.5   Thruster emergency stop switch.

FIGURE 8.6 Main propeller emergency stop switch.

- Availability of power to the thruster motor.
- Thruster motor short circuit.
- Lubricating oil check for pressure and temperature.
- Operational status of thruster.
- Thruster cooling arrangements where applicable (Figure 8.7).

Various alarm systems are designed and installed for the safety of the thrusters on board. For example, alarms and warnings in cases of hydraulic leakage, detection of motor overload, high temperature and high-pressure alarms are designed to initiate audible and visual indicators on the DP control station. Whenever such an event takes place, arrangements should be in place to record them, which can be analysed at a later stage.

Most DP systems require the following signals from the thrusters to be connected to the DP controller or process station. A thruster "running" signal indicating the thruster is running. A "ready" signal is an indicator to the DP system that the thruster is now ready for DP use. Once the "running" and "ready" signals are

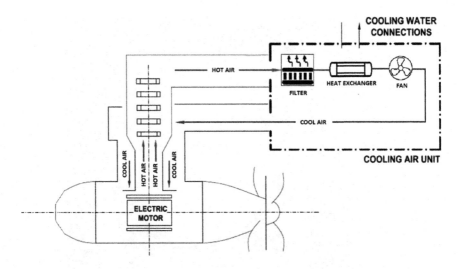

FIGURE 8.7   Thruster cooling arrangement – an example.

available only then may the DP operator be able to start using the thruster by clicking the "enable" button.

All thrusters are designed to receive a "command" signal from the DP controller – this is the TAL. On receiving this command, the thruster is expected to act appropriately, and this is indicated back to the DP controller as "thruster feedback" as an input to the controller. The command and feedback signals are analogue signals, whereas the ready, running and enabled signals are digital signals.

The health of a thruster is easily monitored by having a close look at the command and feedback signals. If the there is a wide gap between command and feedback signal displays on the appropriate views on the DP operating stations/workstations it may be understood that either the thrusters are sluggish or have failed.

For example, if a Class 2 DP vessel is fitted with two electrically driven bow tunnel thrusters, during a DP operation, one of these thrusters fails to thrust fully. Under such circumstances, the DP officer's first action should be to manage the situation by activating the emergency stop for the failed thruster.

In a DP vessel the thrusters may become ineffective when another vessel is operating in very close proximity, the thruster wash from the other vessel will be a cause of disturbance and may increase the DP current/residual as current. When two or more azimuth thrusters are fitted to a vessel, there may be "prohibited azimuth zones" or a "barred zone". This is activated to prevent the azimuth thrusters from interfering with each other.

9 Power Management System (PMS)

9.1 INTRODUCTION

All ships require a constant supply of electricity to perform safe operations. Electrical power is used by electrical drives, auxiliaries, deck machinery, lighting, air conditioning, ventilation (hotel load) and other equipment. Most marine vessels today utilise sturdy synchronous generators to produce the required electrical power. The capacity of these generators will vary depending on the design and intended used of the ship (Figure 9.1).

While at sea many vessels may make use of shaft generators. The shaft generators are powered by the main engine running at a constant RPM. It is important for engineers and others working on modern marine vessels to understand the need to have a PMS on board a ship, its working principles and the design of PMS for a specific type of vessel. Reliable and optimal functioning of the PMS is important to ensure safe, reliable and economical operation of a DP vessel. Hence, it is important that the design premise for a PMS should be well understood by all engineering and operations personnel on board.

Every industrial process needs a dependable, continuous supply of electrical power. The offshore industry deploying different types of DP vessels including DP drill-ships are not immune to this important concern. Whether old or modern dynamically positioned vessels, many of them may use variable speed drives for the ship's propulsion system. Variable speed AC drives commonly known as variable frequency drives (VFD) may also be used for other important operations like drilling, cranes and winches, in addition to a thruster and propulsion system. As VFDs consume quite high power depending upon their design and use, a robust, dependable and well-managed power source is critical for the successful operation of all electrically driven/operated systems. The following figure depicts various signals associated with a VFD (Figure 9.2).

Estimation of DP power requirements is very important for an efficient PMS. The power requirements for maintaining a vessel in position requires high power to counteract the force vectors acting on the vessel. Thus, power requirements are calculated by analysing the configuration of the thrust requirement which depends on the environmental forces acting on the ship. The wind, the waves and the current forces are calculated to arrive at the total thrust requirements to keep the vessel in position. The total thrust requirements become the basis of the calculation of power requirements.

Having understood the power requirements and the status of the power plant, DP professionals may be in a better position to take appropriate action to ensure power is always available for the intended operations. This ultimately results in avoiding

FIGURE 9.1   Main switchboard.

blackout. Thus, the primary function of the power management system becomes to avoid a blackout.

9.2   ESTIMATION OF DP POWER REQUIREMENTS

It is not easy to arrive at the total required power which might be needed to keep a vessel in position fighting against the environmental forces acting on the vessel. The DP controller based on the mathematical model of the ship analyses the forces like wind, waves and current (Figure 9.3).

A DP system is a computer-controlled system used to control the ship's position automatically by using active thrust. Active thrust is the thrust developed by the available thrusters on board.

All ships and floating structures have six degrees of freedom. Three of these movements are on the horizontal plane. These are yaw, surge and sway. The other three movements are on the vertical plane and these are known as roll, pitch and heave.

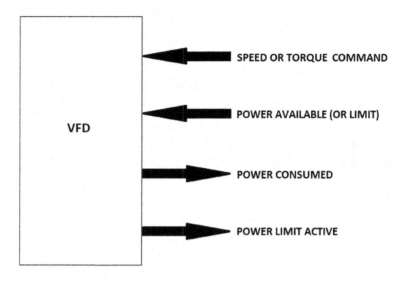

FIGURE 9.2   VFD signals.

All vessels at sea are subjected to forces from wind, waves and current. The total sum of these forces is represented by the force vector. If the vessel must stay and maintain her position the force vector must be compensated by thrust vectors. The DP system uses thrust allocation logic to generate the required thrust from the available thrusters.

It is not necessarily true for all DP vessels, but it is commonly agreed that all DP operations can be conducted by using approximately 80% of total installed power in wind speeds of up to 61 knots. Usually the wind speed is lower than this, hence, the total power used will be much lower.

9.3   POWER MANAGEMENT SYSTEM

A power management system (PMS) for a conventional DP vessel is the system that efficiently matches the need of the power for the existing conditions with the currently running/online generators. If more generators are required online, the PMS can do so efficiently and quickly.

Most DP and diesel-electric powered vessels should have sufficient generators to match the power needs of the vessel. These generators are connected to the switchboard/s as design demands and thus supply power to the motors/drives. The emphasis is to ensure maintaining and improving economy of operations (Figure 9.4).

9.4   POWER BLACKOUTS – CAUSES

Power blackouts on board a ship may take place when electrical demand from various systems exceeds the available capacity. Almost all vessels will have facilities to control both electrical supply and demand. All DP vessels are generally designed to

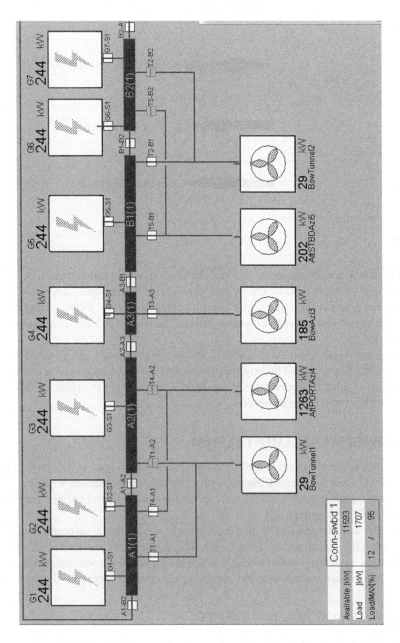

FIGURE 9.3 Power consumption and power available/spinning reserve (courtesy Kongsberg).

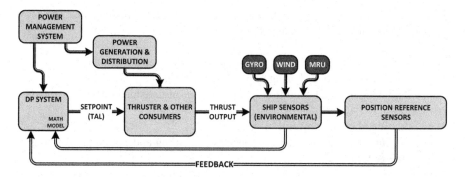

FIGURE 9.4 Power management system and DP system interaction.

have sufficient electrical power capacity to meet the different levels of power demand during different environmental conditions.

If a vessel must avoid a blackout situation then it must be ensured that the power demands must never exceed power availability. The available power is supplied to the switchboard by the online generators.

9.5 POWER BLACKOUT AVOIDANCE STRATEGIES

It is common practice to reduce the non-essential loads from the main power system when the demands increase, so that a complete blackout can be avoided. Pre-set limits are calculated based on the system capacity. One option available to achieve the above is to switch off loads which are already identified. This may help in managing the loads and avoid blackout. Switching off loads is generally carried out by tripping the designated supply breakers.

There are different types of ships carrying out various activities using the power available. Switching off loads may not be safe and suitable for a drillship or other offshore DP-enabled vessels because the fact that the major part of the electrical load is the result of the use of power by the thrusters online.

It is good to note that modern ships are installed with the thrusters and other systems which are controlled by VFDs. Such equipment and drives are solid-state frequency conversion systems, which can be easily controlled by the PMS. For these kinds of loads, control signals can significantly reduce the input power demand of a drive. This arrangement gives better control of load and hence management of power supply on board.

9.6 FUNCTIONS OF PMS

The designed specifications of a power system on board a ship will decide what functions the PMS must have to manage the power. Broadly, the functions of the PMS on board are categorised as primary and secondary functions. The primary function of any PMS is always to ensure power availability. The secondary function is to restore the power within the earliest time defined once a blackout is detected. The most common functions of PMS are as follows:

- Monitoring of generator sets in response to measured parameters.
- Controlling and load sharing of generators online.
- Monitoring the load conditions and thereafter initiate load dependent start/ stop.
- Initiating blackout prevention steps.
- Reducing the consumer loads to avoid overloading and eventually tripping of the generator/s.
- Monitoring electrical power generation and distribution systems so that appropriate actions could be initiated to avoid blackout.
- Monitoring and opening of bus ties if needed.
- Restarting the power system after a blackout is an important function of any power management system.
- For DP-enabled vessels it is very important that sufficient power for thrusters is maintained at all times to ensure that the vessel maintains position.

To sum it up, it may be observed that a marine PMS provides control, monitoring, safety functions and above all a method to avoid blackout.

The PMS has access to the entire power plant, generators, distribution system and consumers.

The visuals provide an overall view so that the power network can be monitored and the generators, switchboards and circuit breakers can be operated. Most of the power consumers are also shown in the graphic representation of the power plant. There may be a way provided to operate it from other applications to maintain continuity of power to various consumers.

9.6.1 User Interface

The user interface for any PMS application depends upon how the electrical power production, distribution and consumers relate to each other. The software will include a display of the each of these including the prime mover and may include the status of the various breakers, bus tie breakers, switchboards and consumers.

It is important to note for all DP professionals that almost all the necessary operations of the PMS may easily be performed from the power management images. In most vessels it is operated by the engine room watchkeeper and in some special cases it may be performed by a DP operator on the watch who by then must have been trained adequately to perform the required functions of the PMS. A few more advanced PMS may have a detailed information about the "status" and "conditioning" of the generator engine and may also include details like exhaust temperature. In the electric generator process section, an image of the power and voltage measurements and related alarms will be shown. This helps the operators/watchkeepers to have a better understanding of the power system.

Many PMS may have a switchboard process image and it may contain various modes selected. This may also have provisions like "load dependent start" and "load dependent stop".

- Generator control operation: Depending upon the level and sophistication of the PMS, the level of automation the degree of the controls may be accordingly designed and implemented.
- Standby mode: This specific setting/mode, will be the defining factor if a generator can be set for automatic start, stop, on load/connect, share load and disconnect functions as required.
- Stop/start mode: Depending upon the design specification, this mode may be provided to help start/stop manually by the operator or automatically with the help of logics in the software. The "not standby" generators may use the manual mode to "stop" but only after making sure that the generator is manually disconnected.

    If a "not standby" generator started, this generator must be on load for any further commands by the PMS.

    Only a "standby" generator is activated automatically. While stopping a standby generator the logic will automatically disconnect the concerned generator after reducing the load.

    Most shaft generators are not likely to have this function.
- Connect or disconnect mode:

    Connecting or disconnecting a generator also can be done either manually or automatically by the system logics. It is important to note here that if a "not standby" generator is disconnected manually, it will not stop automatically. The operator needs to do it manually. While such commands on a standby generator, will make the generator concerned disconnect from load and automatically stop. All concerned must be aware that any disconnection is only possible if there is enough power connected to the switchboard for this function to be done. If a particular generator is on standby mode, it is now programmed to be automatically started, synchronised and connected to the switchboard and share load according to software logics.

9.7   LOAD SHARING

Depending on the needs and the design of the PMS the following modes may be available:

- Manual mode: Generally used while only one is generator in use.

When two or more generators are running in parallel the modes may be as follows:

- Automatic load distribution control: Logically controlled according to requirement and availability.

- Fixed Load: This mode not recommended to be selected for a standby generator.
- Symmetric: This mode of load sharing is one of the automatic modes for controlling distribution of a generator. To come out of this mode you may either select "fixed load" mode, "manual" or "asymmetric".
- Asymmetric: This mode of load sharing is a valuable operating mode especially when you may need to control the load on an engine due to technical reasons such as repairs/after repairs. It is important to note that the "asymmetric" command is not available for shaft generators.

It should be kept in mind that selecting the symmetric command for an incoming generator may affect the load distributed to other online generators in the symmetric or asymmetric mode. Also, if you select manual mode for a standby generator, this will make the generator "not standby".

In a modern diesel-electric DP vessel, power may be generated as follows:

Six or more diesel generators generally fitted in two separate machinery spaces to meet the class requirements. These generators are high voltage generators most of the time, according to the vessel load requirements and machinery installed. These generators are then connected to a split bus system, also called switchboards. Switchboards may be isolated using "bus tie breakers". A "closed" bus tie breaker ensures connectivity of both parts of the switchboard.

The switchboard is the heart and thus is an essential part of a diesel-electric propulsion system of a DP-enabled vessel. All alternators/generators feed the switchboard as per the power requirements and the switchboard then distribute it according to the need of the consumers. An appropriate setup of this equipment is necessary in order to have a workable DP class operation. A bus-tie on a DP vessel is compliant with the FMECA and other design features. This requires an understanding of IMO/MSC 645 and IMO/MSC 1580.

9.7.1 POWER LIMITS

By using this window/setting, one can set the maximum allowed power consumption for major consumers, thrusters and propellers. This will, in turn, affect the power production and consumption of power.

When power consumption/production within these limits is not enough for system operation, an alarm appears, and a set of actions may be initiated by the PMS software to start the standby generator. Power limits are usually set in percentages of maximum consumed/produced power.

9.8 UNINTERRUPTIBLE POWER SUPPLY

DP control systems consisting of DP controllers, hardware (HMI), environmental sensors and position reference sensors (PRS) need a dependable and stable power supply. Often power fluctuations may cause a fuse to blow causing serious damage to sensitive electronic equipment (Figure 9.5).

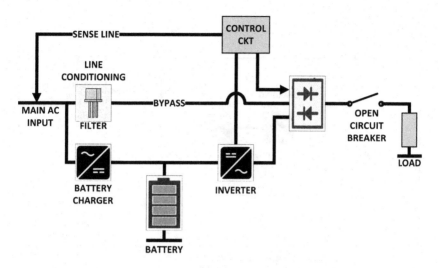

FIGURE 9.5 Uninterruptible power supply.

Thus, most DP electronic components must have backup from a secondary (battery) power source. These battery backup systems are called uninterruptible power supply (UPS). Depending upon redundancy requirements and DP class, a ship may be installed with the required numbers of UPS units. DP Class 2 and 3 must have redundant UPS. All installed UPS on board a DP vessel must last for at least 30 minutes and should be tested regularly (Figure 9.6).

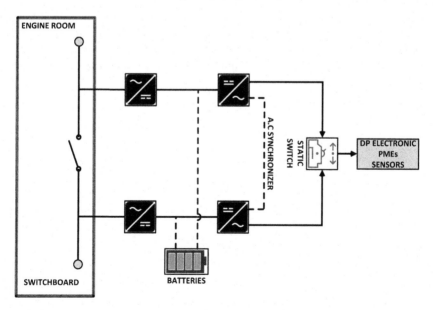

FIGURE 9.6 Non-redundant uninterruptible power supply.

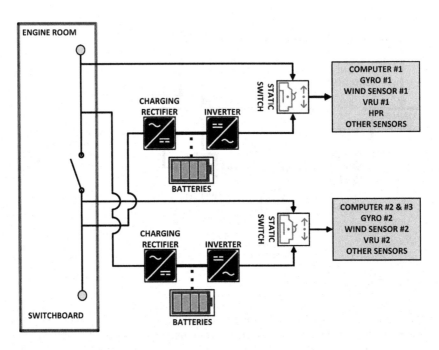

FIGURE 9.7   Redundant uninterruptible power supply.

A non-redundant uninterruptible power supply as shown above is usually suitable for a DP 1 system.

A redundant uninterruptible power supply has special arrangements whereby in the event of one section of a UPS failing, the other section continues to supply power to a dedicated set of equipment and does not result in a control supply blackout as may be the case in a non-redundant uninterruptible power supply (Figure 9.7).

10 Harmonics in Power System

10.1 INTRODUCTION

This chapter discusses how harmonics are generated in the electrical power system of vessels and how these affect the electrical systems on board. The chapter also highlights the measures that could be taken to mitigate the effects of the harmonics.

10.1.1 What are Harmonics?

According to whatis.com: "A harmonic is a signal or wave whose frequency is an integral (whole-number) multiple of the frequency of some reference signal or wave. The term can also refer to the ratio of the frequency of such a signal or wave to the frequency of the reference signal or wave" (Figure 10.1).

Harmonics travelling through power distribution networks may be defined as the disturbances which affect the flow of electricity. As a result of harmonics, electrical power quality is deteriorated and thus the electrical power system efficiency also goes down.

Harmonics in the power system may bring some risks as mentioned below:

- As the root mean square (RMS) currents increase due to harmonics, over-loading of the system may occur.
- Neutral conductors may be overloaded, sometimes exceeding the phase current.
- The humming of a transformer may result in overloading, extra vibrations and early aging.
- The presence of harmonics may also overload the power factor correction capacitors, which may have a long-lasting effect on the ship's power system.
- Sensitive loads are more affected by the presence of harmonics.
- Harmonics also affect communication networks and telephone lines.

In both AC and DC circuits, the resistance characteristics are similar as described by Ohm's law as resistance is a linear device, and the linear devices have a sine wave voltage.

10.1.2 Current Harmonics

Other than the resistive loads, which are considered nonlinear and likely to contribute to current harmonics in the power system on board, a rectifier, which is a

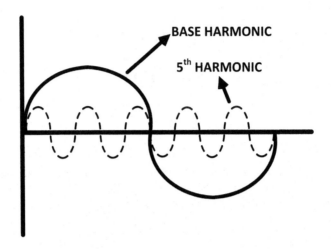

FIGURE 10.1   What is harmonics in power?

nonlinear load, when connected to the power system may result in some complex waveforms which may not be sinusoidal in nature and may cause disturbance. More such equipment fitted on board and may contribute to a complex sine wave, for examples computers, general lighting and battery chargers. One very important contributor to these disturbances is the variable speed drive.

10.1.3   VOLTAGE HARMONICS

Once there are current harmonics in a power system, the voltage harmonics are generated as the voltage supply from the source is distorted because of the impedance. Small impedance may result in smaller voltage harmonics (Figure 10.2).

The complete system as shown above in a single line diagram may be better understood if it is divided into separate parts as explained below:

- One part of the circuit indicating the flow of current at the base frequency.
- The other part of the circuit representing the flow of currents due to harmonic disturbances.

The harmonic voltage flowing through a circuit may be calculated by using Ohm's law.

If a harmonic current of value h flows through impedance Zh, it will result in a harmonic voltage Vh.

Where $Vh = Zh \times Ih$ (using Ohm's law).

It is important to note here that the voltage appearing at point B in the figure above is distorted. This also implies that all voltages supplied through point B will be distorted. The equation above indicates that for a given harmonic current, the voltage distortion is proportional to the impedance in the distribution network.

Some of the interesting terms associated with the harmonics are, odd harmonics, even harmonics, negative sequence and triplen harmonics. Experts believe that even

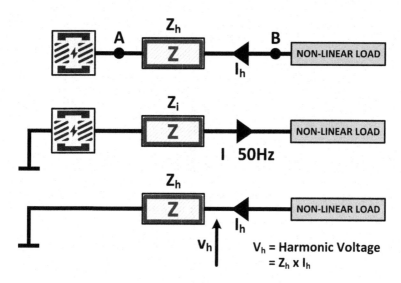

FIGURE 10.2   Current and voltage harmonic.

harmonics are not common. However, for the last few years many cases of even harmonics have been observed in various systems.

Harmonics may be numbered starting from zero for the direct current (DC) component. The No. 1 harmonic is also referred to as the fundamental or base harmonic frequency. This frequency will be either 50 or 60 Hz.

10.2   BASE FREQUENCY

Usually when referring to an AC current or voltage in electrical circuits, it is assumed that these are pure AC and have a sinusoidal shape. This results in a frequency value and this is referred to as the base frequency or fundamental frequency for example 50/60 Hz.

The electrical load on board the DP-enabled vessels, the majority of which are electrically propelled, are not linear. This means that the load does not draw sinusoidal currents, and hence the propulsion equipment load currents will result in a distorted voltage. This distortion is called the harmonic distortion (Table 10.1).

10.3   WHAT IS "TRIPLEN" IN HARMONICS?

The odd and even terminology as familiar with you means two, four, six, eight, etc. will represent even harmonics and the odds are three, five, seven, etc. Harmonics that can be evenly divided by three are commonly referred to as the "triplen".

So, it can be observed that the harmonic number/s three, six, nine, etc., are called triplen harmonics. These two terms of odd and triplen are seen often being used together for these harmonic frequencies. The harmonics may often be classified as per the sequencing they follow, negative, positive or zero.

TABLE 10.1
Various Harmonics and Their Orientation

Frequency Name/No	Base Frequency	2nd Harmonic	3rd Harmonic	4th Harmonic	5th Harmonic	6th Harmonic	7th Harmonic	8th Harmonic	9th Harmonic
Hz	60	120	180	240	300	360	420	480	540
Sequence	+	−	0	+	−	0	+	−	0

10.3.1 Sequencing pattern of harmonics

As can be noted above, the fundamental is a positive-sequence harmonic; the second frequency 2f is a negative sequence and third frequency 3f is a zero sequence. The sequence then continues for the fourth being positive, fifth negative and so on. Positive (+) sequence in forward direction rotation, creates excessive heating effect. Whereas, negative (−) sequence in reverse direction of rotation, will result in motor torque problems. In case of zero sequence, no rotation, there may be heating problems observed due to added voltage/current, or both, in the neutral line.

Due to the fact that the odd frequencies are more common, the triplen odds attract attention for causing overheating of neutrals.

10.4 EFFECTS OF HARMONICS

Almost all power switching devices and solid-state electronic switches may result in cutting and chopping of the sinusoidal waveform of the power supply.

The current drawn is only during the peak of AC supply. This resultant waveform is not sinusoidal and these resultant waveforms may include harmonics which are multiples of the base frequency such as 2xf, 3xf and so on.

For example, if there is a system on board where the base frequency is 60 Hz then 120 Hz will be the second harmonics and 180 Hz will be the third harmonics.

A harmonics sequence may be defined as the phase rotation of the harmonic voltage or the harmonic current as compared to the base harmonics or the fundamental harmonics of a three-phase system. The commonly agreed bad effects of harmonics are excessive current flow in the neutral line and the overheating of the equipment.

Experience of the researchers in this field has indicated that positive sequence harmonics are undesirable as these may result in the overheating of machinery and its components. This is due to the addition of these waveforms of harmonics with the base harmonic.

At the same time, it is not to be assumed that the negative sequence harmonics are not harmful. The negative harmonics circulate between the phases and this may result in additional problems with motors/drives. This may weaken the magnetic field required by motors to meet the load requirements and the loss in torque may result in further problems.

Currents in the positive and negative sequence harmonic are arranged in such a way that they cancel out each other. The triplen harmonics do not cancel out and thus add up arithmetically for the neutral wire which is common to all phases.

This may result in the triplen current in these harmonics to rise up to three times. This high current may ultimately result in overheating and thereby reducing the efficiency of the motor or drive system.

It may be interesting to summarise the effects of the harmonics as follows:

- Distortion of currents and supply voltage waveforms is not good for the ship's power system.
- This could cause overheating of transformers and rotating equipment due to increased power dissipation in the electrical machinery.

- The insulating material of windings, etc. may age faster and deteriorate.
- The life cycle of the equipment may be reduced.
- Unexplained overloading of electrical equipment.
- Overheating of electrical/electronic equipment may cause malfunctioning.
- Hysteresis losses may go high.
- The real power (kVA capacity) of the machinery may decrease.
- Overloading of neutral line may cause risk of fire.
- Increased neutral to ground voltage levels.
- Harmonics may cause capacitor banks failure.
- Harmonics may also cause unexplained tripping/failure of starters, breakers and fuses.
- False/erroneous reading by meters, measuring devices, etc.
- The above may require increased maintenance and hence the cost of operation will go up.

10.5   FINDING AND MEASURING HARMONICS IN POWER SYSTEM

A complex waveform is one which may contain many other harmonics which are in addition to the base harmonics. These could be in phase with the base harmonics or may be out of phase.

The waveform appearance will start becoming more complex as the spectrum of harmonics becomes more cluttered and complex.

This indicates that the resultant sine wave is not an ideal sinusoid.

Harmonic measurements are carried out for the following reasons:

- For network mapping, to get an overview of the power distribution.
- Corrective measures may be planned to make out the origin of a disturbance which may be causing the harmonics. This is helpful to eliminate the harmonics present.
- To check if the corrective measures initiated are effective to keep the harmonic disturbances at generally accepted levels.
- In a shipboard power system, it is important to know the best way to measure the harmonics. The below points given may be helpful:
- An expert engineer may be sent to witness the site for a defined period so that the expert can suggest his/her perceptions.
- The harmonics may also be measured by instrumentation/measuring devices installed for the purpose (temporarily). This may be done for a longer period and may give more reliable results.
- Harmonics measurement may also be done by special devices permanently installed in the power system.

10.6 CORRECTING HARMONICS IN POWER SYSTEM

It is generally agreed that major loads like the electric propulsion system can be important causes of harmonic disturbances in the power system on board, as modern electrical propulsion utilises variable frequency drives (VFDs). When vessels are being designed with electrical propulsion and other major electrical loads, it is imperative that the harmonics mitigation steps are also included to ensure reduction of the resulting distortion in the ships power system.

Industry practices, existing rules and regulations formulate guidelines to deal with the harmonics in power systems so that their effects can be mitigated.

Accordingly, the designers and yards calculate to arrive at these distortions so that the effects on the power plant of the ship and other machinery being adversely affected can be reduced.

Corrective measures are taken in case such disturbances are observed. For this purpose, measurements of voltage and current are carried out:

- At the source level of power supply.
- On the distribution switchboard and, also on the main consumer switchboard.
- Also, check at the outgoing of each distribution switchboard.

For good results, it is recommended to observe for a full work cycle so that all the variations are recorded. Looking at the deteriorating effects of power harmonics, the IEEE developed standard 519-1992 these standards define recommended general practices and measures to be taken to control harmonics.

11 Introduction to Environmental Sensors

11.1 INTRODUCTION

This chapter outlines the basic need to have environmental sensors used in dynamic positioning (DP) systems, their working principles and the common troubles associated with these sensors. The main environmental sensors used in a DP system are the wind sensor, gyro compass, vertical reference unit (VRU)/vertical reference sensor (VRS)/motion reference unit (MRU).

11.2 SENSOR REDUNDANCY

Sensor redundancy is essential for maintaining the desired class of operation and thereby safety of operations.

11.2.1 SENSOR REDUNDANCY IN TRIPLE SYSTEMS

The normal practice in a triple-redundant system to achieve true redundancy is to use at least three of each type of environmental sensor. The triple-redundancy concept for environmental sensors can be understood better with reference to the diagram given below. This diagram, using an example of three gyros, explains how the concept is helpful. The same concept is extended to all other sensors to achieve triple redundancy. As shown in the figure below, triple-redundant environmental sensors, i.e., gyro, wind sensor and MRU are enabled, but the preference is given to Sensor 1 in all cases (Figure 11.1).

11.2.2 SENSOR REDUNDANCY IN DUAL SYSTEMS

Each sensor is interfaced with both controllers. Both the computers take input from all sensors and also share with each other. When there is a failure detected within one of the computers, due to interface/card/input/output devices failure, this computer will start using readings from the other computer. In this case an alarm is generated. Also, with dual-redundant sensors, the DP system cannot perform first-level voting, but a comparison of the data is carried out.

11.2.3 SENSOR REDUNDANCY IN SINGLE SYSTEMS

A single-sensor system cannot perform first level voting of sensors, but second level of voting is done. A warning on rejection of a sensor will be given to the operator. The second level of voting detects errors.

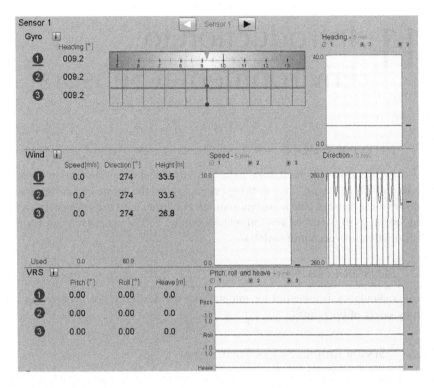

FIGURE 11.1   Environmental sensors (courtesy Kongsberg).

11.3   VRU/VRS/MRU

By now it is well understood that the DP system maintains a vessel's position by controlling surge and sway, and heading is controlled by controlling the yaw movement. Heave, roll and pitch are monitored to enhance position data received from position measuring equipment (PMEs) or position reference sensors (PRSs). To measure roll, pitch and heave we can make use of VRU/MRU as suitable.

This also it is a well-known fact that the vessel, while at sea, facing various environmental forces acting upon it, will eventually result into some movement. As the vessel rolls or pitches, positions of PMEs may be due to these being fitted away from the centre of gravity (CG) of the vessel. If not configured properly this offset may also be understood by the DP system as an actual movement of the vessel. The environmental sensors (VRU/VRS/MRU) are used to measure these movements and applied as an input to the controller in order to offset the movement of reference sensors.

Some specialised vessels may use this input for the stabilisers to compensate the vessel's movements using stabilisers.

A VRU/VRS/MRU is fitted in the DP vessel to measure pitch, roll and heave. VRU and VRS are used interchangeably and measure roll and pitch, whereas MRU measures roll, pitch and heave. From the DP perspective, the effects of pitch and roll are more critical to position keeping.

A VRS/VRU in its most simple configuration, will have a damped pendulum floating in a chamber containing fluid. The detector coils inside convert the position of the pendulum to an analogue voltage. An advanced VRU/VRS may also have facilities to measure heave. MRUs are designed to measure all three movements of the vessel on the vertical plane, i.e., the roll, the pitch and the heave. An MRU, on the other hand, works on the principle of linear accelerometers to measure accelerations and this information is then used to calculate inclination angles. This signal produced by the VRU/VRS/MRU then may be used to represent angles of roll, pitch and heave motions of the ship.

11.4 GYRO COMPASS IN A DP SYSTEM

A gyro compass provides heading and rate of turn data to the DP system. The number of gyro compasses installed would primarily depend upon the DP class and the utilisation of the vessels.

Depending on the requirement, if there are two gyro compasses installed, the DP system is designed to monitor the difference in heading data. If there is a difference detected, a warning is issued when the difference exceeds a set value. These aspects are usually tried out during the initial setting of the equipment and the DP system (Figures 11.2 to 11.4).

In case there is a need to have three gyros, the DP system can make use of voting and determine a gyro failure. Accordingly, this can be used to initiate a warning. Different makes of gyros are available in the market to suit the requirement according to the design of the vessel. Most DP 2 and DP 3 vessels are fitted with three gyros.

New technological advancements are bringing in new methods to provide the required heading input to the DP system. One such method is differential global positioning system (DGPS) using strategically positioned DGPS receivers and motion sensors.

11.4.1 GYRO COMPASS FAULTS

Sometimes all three gyros fail at the same time which can be a nightmare for the crew. So, it is always a good idea to have adequate knowledge about the common failure modes and faults on gyro compasses. Some of them are listed below:

- On certain occasions, it may be seen that a gyro may take a long time to settle after power-up. This may be after a blackout or shutdown. To protect against the effects of a blackout on such important equipment, 24 V DC power is normally backed up to 220 V AC power supply.
- Gyro compass error may be due to a wrong latitude setting. This error may be easily corrected by setting the correct latitude manually. It is good to remember that most new gyro compasses have an input from the GPS which can feed latitude input directly.

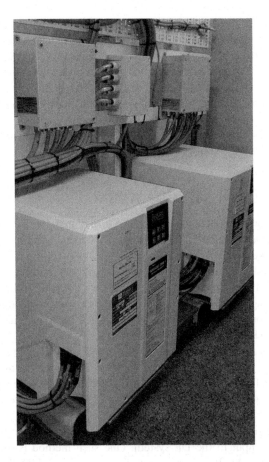

FIGURE 11.2   Gyro compass units.

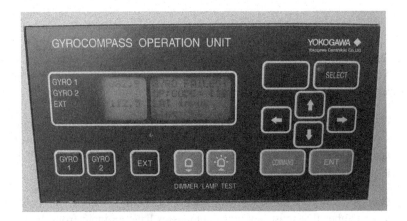

FIGURE 11.3   Gyro control/operator unit.

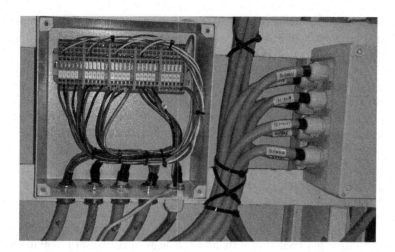

FIGURE 11.4   Gyro signal distribution panel.

- It is normal practice to support a gyro compass with both an AC supply and a backup 24 V DC supply. This helps to maintain proper functioning even after power failure. Gyro heading input may be utilised for error correction of the position as the gyro input is connected to most of the PRS. Gyros may require latitude correction to be in manual setting due to certain requirements. This may result in errors if no proper settings are done. To avoid this, some DP vessels are using fibre optic or laser ring gyros as these do not need latitude correction.
- Any failure of PRS affects the DP control system. All PRS failures may result in no control of surge and sway movements. However, as the gyro is able to input the heading and rate of turn (ROT) signals, it is still possible to control the heading (yaw) movement of the ship.

11.5   WIND SENSORS

Wind has the potential to blow a vessel off position. Therefore, DP systems need wind speed and direction data from wind sensors to compute the effects of the wind on the vessel's superstructure and hull. This will be used to determine the thruster force required to counteract the effects of wind on the vessel. The information is also used to arrive at the weathervane or minimum power heading mode which is very common in shuttle tanker operations in offshore loading.

11.5.1   Types of Wind Sensors

Rotating cup anemometer: Several types of wind sensors are fitted aboard vessels. Generally, a wind sensor may have a set of rotating cups which rotate as the wind blows on the cups. The voltage generated will precisely indicate the speed of the wind on an appropriate indicator.

Impeller attached weathervane: Another type of wind sensor has the impeller attached to the windvane. The rest of the functions are the same as the rotating cup anemometer (Figure 11.5).

Ultrasonic wind sensor: This wind speed and direction measuring equipment uses a fixed set of ultrasonic transmitters. The time delay measured can be utilised to calculate the speed and direction of wind.

These ultrasonic type wind sensors require literally very little or no maintenance, hence they provide long service even in harsh weather conditions (Figure 11.6).

11.5.2 Wind Feed Forward Function

The wind sensor provides an input to the mathematical model. Due to the time taken for the mathematical model to build up the process, the mathematical model takes time to evaluate and thereafter respond to changes in the vessel or environmental forces acting on the vessel.

The wind can suddenly change its speed or direction or both. This will have a corresponding effect on the position-keeping capability during these sudden changes. To avoid this, a wind sensor is also connected to the DP system directly without going through the mathematical model. This is known as the "feed forward" function.

The feed forward function helps the DP system to react immediately to these sudden changes in wind forces acting on the vessel. To avoid wind feed forward initiating heavy thrust during helicopter operation around the vessel, the wind sensors are deselected.

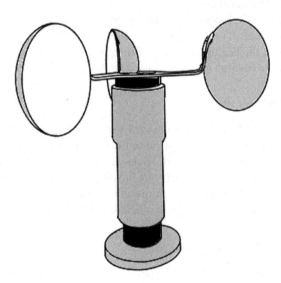

FIGURE 11.5 Wind sensor – cup type.

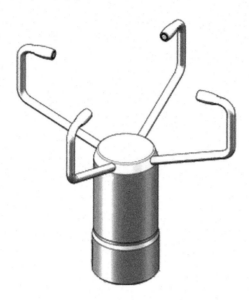

FIGURE 11.6 Wind sensor – ultrasonic type.

12 Introduction to Position Reference Sensors

12.1 INTRODUCTION

Dynamic positioning (DP) systems require reliable, accurate and continuous input from the position reference system (PRS) or position measuring equipment (PME). Some DP operations may require better than meter-level accuracy, which depends upon the input from a dependable PRS. The data update rate for a DP system is once per second to achieve high accuracy. As offshore operations are carried out under stricter safety requirements, reliability is a very important factor, and this reliability of positioning depends upon the reference sensors.

The PME/PRS are independent of a ship's normal navigational equipment. The basic purpose of the PRS/PME is to provide reliable and continuous position reference to the DP system. The sensors may work on different working principles, but they all are designed to provide the DP system with the position reference.

12.2 BASIC WORKING PRINCIPLES OF THE PRS/PME

1. Radar or microwave.
2. Laser.
3. Hydro-acoustic.
4. Mechanical.
5. Frequency modulated continuous wave radar (FMCW radar.)

12.3 RADAR/MICROWAVE-BASED SYSTEMS

The leading radar/microwave-based position reference systems currently in use are Artemis and the Global Navigation Satellite System (GNSS). The Artemis is simply a low power 3 cm radar whereas the GNSS uses signals sent from satellites to calculate the position.

12.3.1 ARTEMIS

The Artemis system, made by Wärtsilä Guidance Marine, UK, is based upon the microwave principle and is a 9 GHz (3 cm) radar. This PRS is used for special long-range applications, particularly for offshore loading. One Artemis is installed on the target vessel floating production storage and offloading (FPSO) and a second Artemis on the approach vessel (shuttle tanker).

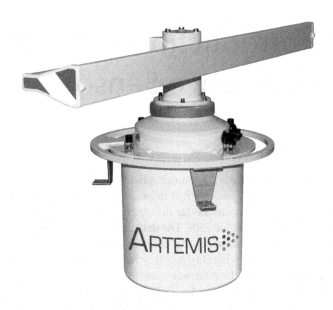

FIGURE 12.1   Artemis Mk6 sensor (courtesy Wärtsilä Guidance Marine).

The position of the approach vessel is calculated by arriving at the distance in absolute terms and the angle between the two Artemis antennae. Being a 3 cm radar, the equipment range is the maximum among all the DP sensors, and the latest MK 6 Artemis has an operation range of 10,000 metres. The Artemis Validator is a diagnostic tool that can be used to test and commission an Artemis without the need for a second Artemis (Figure 12.1).

Being made up of two identical radars working in the near vicinity, Artemis provides an all-weather operation for specialised offshore operations. This sensor is known to have very high accuracy and is very easy to use. An Artemis system is made up of three main components which, of course, may vary depending on the system configuration ship to ship. Firstly, it has a fixed station, and this radar is installed on the target vessel, usually the FPSO. Secondly, a mobile station for which the position is determined, is fixed on board a shuttle tanker. Thirdly, to operate the system there is Artemis Dashboard software which is used by the DP operator for setting up and using the system for position keeping of the vessel.

12.3.2   GLOBAL NAVIGATION SATELLITE SYSTEM (GNSS)

All GNSS-based systems and especially the GPS (the U.S. navigation satellite timing and ranging) (NAVSTAR) system has revolutionised the scenario in navigation and positioning services. In addition to the GNSS-based navigation and positioning services and other such devices and systems, there are quite a few other sensors which are being used as sensors for this purpose. Radar is considered one such prominent navigation aid, which, with some adjustments, can also be used as a position reference sensor.

The Global positioning system (GPS) consists of three main components, as they are commonly known, as segments are mentioned below:

- Space segment.
- Control segment.
- User segment (Figure 12.2).

A number of satellites circuiting around the earth in an orbit make the space segment. The number of satellites available for global navigation satellite system (GLONASS) and NAVSTAR are different and the dependability in a way depends on this.

The control segment is located in safe places ashore as this is controlled by the defence departments of the countries concerned. All the GNSS providers may have their control stations at convenient and strategic locations worldwide for better services.

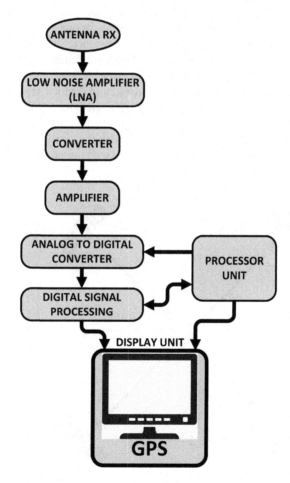

FIGURE 12.2   GPS basic system.

The user segment consists of the hardware and the software associated with the use of these services by any user. All users, including the DP operators and the technical personnel on board, must be familiar with these components.

12.3.3 MEASURING OF POSITION

A receiver on board is constantly listening to or receiving these signals from all the satellites available in the range. The information received and the formula based on the distance, speed and time is used to calculate the distance of the receiver from the satellite. The number of satellites will influence the accuracy and usability of the GNSS equipment. With one satellite in range only the distance between the receiver and the satellite can be measured.

With two satellites in range the position is little better than having one satellite.

When three satellites are in range, with the help of triangulation or trilateration the position received is improved. But the trilateration may have a problem and the position may jump due to the signal becoming weak for one or more of the satellites.

Satellites communicate with receivers, receivers determine distance (by using distance, speed and time formulae) and the distance calculated from multiple satellites is used to calculate the location of the receiver.

The process of measuring distance requires clocks which are very precise and hence atomic clocks are used. These clocks are known to have 0.000000003 seconds accuracy. Onboard receiver clocks are required to be corrected as well. This is achieved by synchronising the receiver clocks with the satellite clocks. If there are a few satellites and they are geometrically not spaced out properly, there will be an error in position measurement (Figure 12.3).

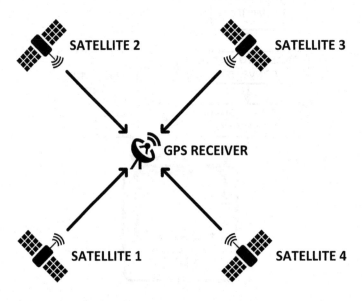

FIGURE 12.3 Number of satellites for GPS.

This makes it important that there are enough satellites, and they are well spaced geometrically in each direction. The minimum number of satellites required is four which can be set to activate an alarm in case the tracked satellites go below the set number.

Terms/terminology and definitions which are quite useful for people involved in the use and maintenance of the GNSS are given below:

- Position dilution of precision (PDOP): PDOP may be defined as an error in GPS. For most applications offshore, a PDOP range of 4–6 is considered appropriate for operations – the lower the better, less than 4 is considered excellent and more than 8 is considered poor.

  It can be used as a tolerance setting for acceptability of GNSS signal quality and may be referred to as a "PDOP mask" or filter.
- Geometric dilution of precision (GDOP): GDOP may be defined as the location estimation of the satellite constellation at a given time and position.

There are other terms and terminology attached to GNSS such as:

- HDOP: Horizontal dilution of precision.
- VDOP: Vertical dilution of precision.
- TDOP: Time dilution of precision.
- RDOP: Relative dilution of precision.

12.3.4   GPS MASKING

GPS masking may be defined as the process of controlling the data quality of the incoming data from satellites. The process ensures that the unwanted data are not allowed to come in.

PDOP masking aims at ensuring that position is recorded only when the satellite geometry is good enough. For example, there may be four satellites with good precision or may be six satellites with reasonable levels of precision or sometimes even eight satellites available with average or not so good precision.

Elevation masking is used to set a minimum elevation angle above the horizon. Only satellites above this angle will be used. This helps to avoid signal diffraction and also avoids the increased chances of multipath errors.

Differential correction is the process used to apply various corrections at the base stations ashore. The corrected signals are then transmitted to the end users by using appropriate methods as suitable for the user.

SNR is the signal to noise ratio. Masking the SNR is used to ensure better reception. The higher the ratio, the better it is considered for operations. The normal range of SNR is between zero and 35. A range between ten and 15 is typical, and below five is not advisable to use.

PDOP $\times$ GDOP = The overall estimation of accuracy of the receiver distance may be calculated as below.

GDOP $\times$ PDOP = Overall accuracy. For example if we have a PDOP of 5 and GDOP of 2.5 metres, then the overall accuracy will be $5 \times 2.5 = 12.5$ metres.

12.3.5 DIFFERENTIAL GPS

The standard GPS signal may not give the required accuracy as required by most DP vessels, so there is a need for better accuracy. This is provided by third party service providers like Veripos and Fugro for example.

The vessel and a shore-based station for providing the differentially corrected signals are expected to be in same area and receiving the signals from the same set of satellites.

The shore-based station applies the corrections required for all errors except the multipath and signal noise errors. The corrected signals are sent to the ships using the most effective method depending upon the location of the ship.

GNSS is designed to operate with a number of satellites in the orbit.

Satellites are commonly used for communication and navigation purposes. It is interesting to know about the different services available and the limitations associated with them. A brief description is given below for each and it is recommended to refer to more details as and when needed.

12.3.5.1 Global Positioning System (GPS)

NAVSTAR (the U.S. system) and GLONAS (the Russian system) are known as global positioning systems (GPS). These two are the most popular and frequently used GNSS. GPS services have been available since the late 1980s, but positioning services were made available for commercial use much later.

Essentially the system is comprised of minimum of 24 satellites placed in an orbit around the earth at a distance of approximately 20,200 kilometres. The Earth has been divided into six zones for this purpose and each zone will have coverage from a minimum of four satellites. Each satellite covers an area in a way to provide the required input for positioning. A minimum of three satellites are required for the purpose of triangulation/trilateration. To ensure accuracy there is a need to have input from four satellites.

The International Maritime Organization (IMO) signed an agreement and accordingly the U.S. department of defence (DoD) will be obliged to give a notice of six years to terminate the services of GPS for civilian use (maritime use). Refer the following link for more details: www.pnt.gov/public/docs/1994/imo 1994.pdf

12.3.5.2 GLObal NAvigation Satellite System (GLONASS)

The GLObal NAvigation Satellite System commonly known as GLONASS was the Soviet Union's answer to the Americans in this field. The development began almost at the same time, but the first operational satellite was launched and tested in 1982.

During the 1990s, due to the disintegration of the Soviet Union, GLONASS operations were disrupted and then the Russian government rededicated the system to users in 2000.

The full operational status of GLONASS requires a number of satellites in use and a few spare satellites . Operating in high latitudes, it needs a minimum of 18 operational satellites to provide round-the-clock positioning services.

Recently, the Russian government has brought the GLONASS system back to almost full operational status by launching additional satellites required for the purpose, but to replace the ageing satellites, this needs to be continued. The GLONASS system has proved to be a good help in augmenting the GPS services during the disturbances like ionosphere, scintillations and poor satellite coverage. The Russian administration is investing heavily to make GLONASS a real and truly independent alternative to GPS.

12.3.5.3 Galileo

The space agency of the European Union has started a new GPS service called Galileo. To make Galileo fully functional, a total of 30 satellites have been planned. A minimum of 27 satellites is needed for a fully functional system with three spares. There have been unexpected delays and a resource crunch in the past and things seem to be moving in a good direction now. Notwithstanding this, Galileo is not considered as a viable alternative to GPS any time soon.

12.3.5.4 BeiDou/Compass

The Chinese system is known as BeiDou and the services are also known as Compass. This system is being used by the Chinese for navigational purposes. The Compass satellites were launched by the Chinese space authority and it is expected to have total 35 satellites once the complete project is up and working. This system is expected to have fully operational status by 2020.

12.3.5.5 The Quasi-Zenith Satellite System

The Japanese system for regional GNSS which uses satellites for navigation purposes is a system known as the Quasi-Zenith Satellite System (QZSS). The system is used for navigational purposes in the Asian Pacific Ocean region and Japan.

12.3.5.6 IRNSS

IRNSS or the Indian Regional Navigation Satellite System is the government of India and its space agency ISRO's answer to the need for satellite navigation for the region. The process of augmentation is progressing well to make IRNSS a full-fledged GNSS.

12.3.5.7 GPS Augmentation Services

Experience shows that a standalone GNSS may not be good enough to provide the kind of accuracy and precision needed for offshore operations. There are some free services and some paid services available to meet various requirements.

These services currently available on the market make use of two different types of augmentation techniques namely, relative and absolute. By utilising the augmentation services the user can avoid erroneous input from these positioning services along with integrity checks.

Augmentation services provide positioning accuracy and integrity check of GNSS. This helps in detecting erroneous signals from being rejected and thereby providing better positioning.

12.3.6 RELATIVE GPS

Differential, Absolute and Relative Position Sensor (DARPS), GPS by Kongsberg)

DP shuttle tankers make use relative GPS when while approaching FPSO for loading. The FPSO may utilise various options to hold position such as weathervane or turret mooring.

When the stern of the FPSO moves due to the above positioning requirements, the position requirements of the shuttle tanker also change and make it a very complicated situation. This complicated situation is understood and overcome by using a relative GPS.

A Differential, Absolute and Relative Position Sensor (DARPS; GPS by Kongsberg) is placed on the FPSO is the equipment to be used to resolve the above complications due to relative dynamic positioning of the shuttle tanker and FPSO.

12.3.6.1 Absolute Techniques

This technique very often called precise point positioning (PPP). This is a comparatively new concept in satellite positioning. For achieving this an arrangement is made to utilise a network of reference stations to check the data and calculate errors which may be present in data received from the satellites.

This final positioning service received is very accurate. To achieve this the end user does not necessarily have to be near a reference station.

12.3.7 CRITERIA FOR SELECTION OF GNSS

While selecting a GNSS (satellite positioning system), it is necessary to look at a number of factors to ensure that existing and near future requirements are met.

It is important that the following points are considered:

- The level of accuracy.
- Level of redundancy.
- Main features of the system.
- Geographic/coverage areas.
- Planned deployment of the vessel.
- Service and support available for the services.
- System integration and failure modes.
- Installation requirements.
- Training requirements.
- After sales support/maintenance.

Advantages of GPS/DGPS:

- GPS/DGPS have high accuracy.
- Easy to deploy and use.
- Satellite constellations available globally.

Limitations of DGPS:

- Proximity to drilling platforms or oil rigs (these structures interfere with satellite signals and differential corrections due to multipath error).
- Accuracy of DGPS/GPS is affected by solar prominences.
- DGPS accuracy deteriorates with increases in distance from reference stations.
- Additional cost for differential correction.
- Global coverage (DGPS have some restrictions in the polar regions).

12.4 LASER

Laser is the short form of Light Amplification by Simulated Emission of Radiation. Laser may be defined as a beam of photons which are coherent and focused. It is coherent as it uses only one wavelength, in comparison to other light waves which may have many wavelengths.

In 1960 Theodore H. Maiman at Hughes Laboratories successfully made the first working laser. This practical experiment was based on the theoretical research of Charles Townes and Arthur Leonard.

The laser-based position reference sensors used in DP systems are listed below:

- Fanbeam.
- Cyscan.
- Spottrack.
- SceneScan.

12.4.1 FANBEAM

Based on laser technology, Fanbeam is a positioning reference sensor which can work with high accuracy to position offshore vessels for carrying out various tasks near a platform or other structures (Figure 12.4).

The system is comprised of a laser sensor installed on board the DP vessel. This sensor has a fan shaped beam which will be able to receive the reflected signal from the reflector placed on the target vessel. The sensors can be tilted to seek and lock on the target.

The software and user interfaces allow selection of either single or multiple targets. By feeding gyro input to the sensor the relative heading can also be obtained, especially while working with a moving target.

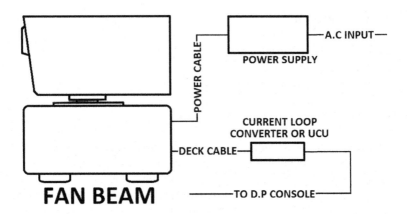

FIGURE 12.4   Fanbeam – laser sensor.

The laser radar is connected by two cables, one for power from the power supply unit (PSU) and the other goes to the universal control unit (UCU). The UCU is then connected to a display for the operator to operate and carry out settings. A serial output to the DP controller is connected from the rear of UCU.

For smooth and accurate operations, it is pertinent that the target and the transmitting and receiving lenses are kept clean.

12.4.2 CYSCAN

The CyScan reference system is made by Wärtsilä Guidance Marine, UK. CyScan is as a local position reference sensor for DP operations. Range and bearing with respect to a reflector target can be determined by CyScan. The targets can be of different types depending upon the job and the position accuracy required. The system also uses automatic wave compensation (Figure 12.5).

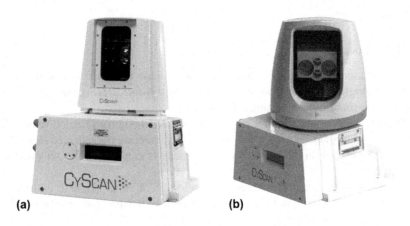

FIGURE 12.5   (a) Cyscan 12.5, (b) Cyscan AS (courtesy Wärtsilä Guidance Marine).

The range of a standard system is up to 2500 m depending upon the type of reflectors used and the prevalent weather conditions. CyScan has a scanning range of 360°. Use of multiple targets can give a heading reference. Cyscan can be interfaced with DP systems using ethernet or serial communication. The "dashboard" user interface provided is used to configure the Cyscan system, its control and handling of data and alarms.

The latest version, Cyscan Absolute Signature (AS), eliminates common issues found with traditional laser systems, such as false reflections, and provides better performance during target identification, acquisition and tracking for DP operations. The CyScan AS sensor works in combination with special retro reflector prisms that have a black tinted prism/lens filter. The CyScan AS sensor can identify the unique signature of these prisms and therefore can ignore any false reflections. When there is no AS reflector available, it works like a normal CyScan.

12.4.3 SpotTrack

This is a new addition to the existing laser-based systems in the market, by Kongsberg. This is a lightweight, stabilised laser sensor, which can offer precision range and bearing readings to either one or multiple targets in DP application. A unique feature, automatic wave motion stabilisation provides better target lock. Once the target is detected and locked, calculations are carried by the SpotTrack sensor and sent to the DP system (Figure 12.6).

12.4.4 Target Reflectors for Laser Sensors

To get the best performances from the laser-based sensors it is advised that the reflectors used are of good quality. Generally, the following types of targets are used:

- Reflective tape (flat) – recommended for commissioning only.
- Cylindrical targets – for close range operations only (<200 m).
- Prism – recommended for all DP operations.

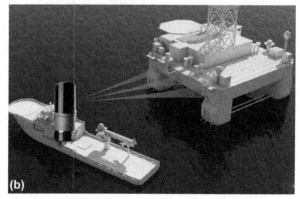

FIGURE 12.6 (a) SpotTrack 12.6 (b) SpotTrack in Use (Courtesy Kongsberg).

The flat targets have limitations for use only from a particular direction compared to cylindrical and prisms. Tape-type targets can work well up to maximum 300 m range depending on the weather and quality of target. Prisms are highly reflective and are recommended in DP operations. Multiple prisms are mounted in a single housing at different angles to give maximum angular coverage. Any angle operations require 360° viewing and this needs a minimum of six prisms. To avoid any potential blind spots, it is recommended to have eight prisms. Prism targets can give an effective range of 2000 to 2500 metres.

12.4.5 SCENESCAN®

SceneScan by Wärtsilä Guidance Marine, UK, is the first targetless position reference sensor working on the laser principle. Rather than tracking individual targets, it tracks off the whole scene. It uses simultaneous location and mapping (SLAM) so that all the current observations in a scene are compared with the previous observations of the scene, so the range bearing and heading to a virtual reference point can be calculated (Figure 12.7).

FIGURE 12.7   SceneScan (Courtesy Wärtsilä Guidance Marine).

SceneScan can be used for almost every offshore structure and gives the vessel complete independence from the asset. Since it does not rely on physical targets mounted on the asset the vessel now has the ability to track off almost any part of the structure at any approach angle or from any direction. SceneScan also has a monopole mode feature for use at offshore wind farms and mitigates the requirement for many reflector targets to be installed on every monopole.

12.5 ACOUSTICS

Acoustics, simply defined, is the interdisciplinary science which deals with the study of all waves. These waves may travel in gases, liquids and solids. The sensors discussed in this category are also referred to as hydro acoustics and the sound wave travels in water.

Sound energy propagates in water at a much higher speed and efficiency than it does in air (Figure 12.8).

Acoustic-based position reference systems have been in use for DP applications for a long time. The time interval of the acoustic signal travelling in water is measured to calculate the ship's position. Similarly, the position of an ROV and other underwater installations can also be calculated. Depending upon the accuracy and redundancy requirements, acoustic systems (HPR/HiPAP) may be categorised as shown below.

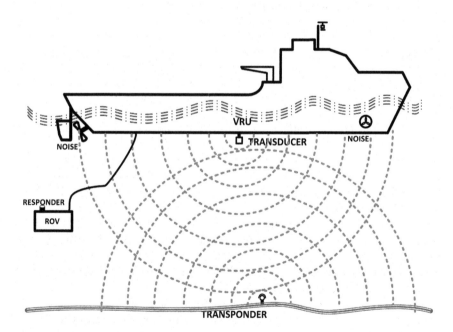

FIGURE 12.8 Hydro acoustic basic system.

12.5.1   Hydroacoustic Position Reference (HPR) or High Precision Acoustic Positioning (HiPAP)

Depending upon the baseline, the HPR/HiPAP system may be divided into the following types:

- Ultra-short baseline system (USBL).
- Short baseline system (SBL).
- Long baseline system (LBL).

The most commonly used position reference sensor for high-end DP systems, HPR or HiPAP for DP position reference are made up of a transducer mounted in the hull of the DP vessel. This transducer transmits an interrogating pulse. This pulse, once received by the respective transponder/s at the seabed or an underwater device, automatically activates one or more transponders positioned on the seabed. This will make the transponder send out a reply.

The transducer/s will then receive the reply. The time difference between transmission and reception is used to calculate the range and angle at the transducer head. Thus, the DP vessel's position relative to the transponder is determined. This position is known as absolute position.

12.5.2   Ultra-Short Baseline or Super-Short Baseline (USBL or SSBL) System

In the USBL/SSBL system, one hull-mounted transducer transmits and receives acoustic messages from transponders deployed on the seabed. For USBL/SSBL one-time calibration after installation is required. Calibration and alignment will be important factors in determining the accuracy of the reference system. Accuracy is limited to approximately 0.25 to 0.5% RMS of slant range from the transponder. As the water depth increases, the accuracy decreases. The system is also adversely affected by the noise generated by thrusters and other work being done in the vicinity.

The direction of the transponder can be calculated by knowing the direction of the incoming signal at the transducer. One time-phase comparisons made between pairs of receiving elements (inside the transducer) are used for this purpose. A Kongsberg SSBL transducer head is normally designed to have 48 elements (Figure 12.9).

To use the system for positioning, the computer sends a command to the transceiver. The transceiver then sends the message to be transmitted using acoustic waves through the transducer head. The transponder below gets activated and sends back a reply to the transducer. Upon receiving the reply, the transceiver measures the time delay and the time-phase data.

These data are processed by the processor to determine the slant range and direction. Roll and pitch information is fed into the computer by the VRU/MRU for position corrections.

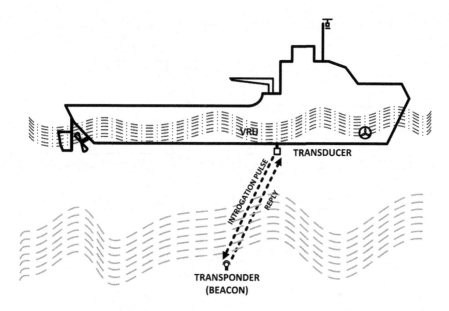

FIGURE 12.9   Ultrashort/supershort baseline system.

Positioning data is displayed in the display system. These data are then sent to the DP controller using appropriate signals. Most systems currently use network signals for communicating with the DP controller.

12.5.3   Short Baseline (SBL) Systems

A short base line system is not a commonly used system. This system makes use of an array of transducers also called as hydrophones, which are installed below the ship. The distances between transducers are used as baselines and it may be the range of 10–50 meters. A transponder positioned on the seabed transmits periodic pulses at a set frequency and at a set interval.

By calculating the time difference between transmission from the pinger and reception at three or more transducers the vessel's position can be calculated (Figure 12.10).

An SBL system has the following advantages:

- Good update rates to the DP system.
- Calibration required only at the installation.
- Better accuracy as compared to USBL. The overall accuracy is approximately 0.15% RMS of water depth of location.
- As the system is pinger-based, interrogation through water may be avoided thereby reducing signal noise one way. This helps in achieving better results.

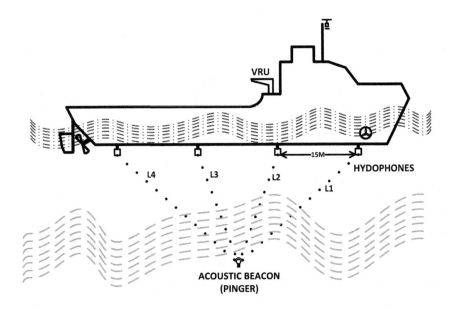

FIGURE 12.10   Short baseline system.

The SBL system also has some disadvantages as mentioned below:

- VRU/MRU calibration affects accuracy.
- Compared with other variations of HPR/HiPAP, this system requires many transducers (minimum of three). This has a direct effect on the cost of the equipment.
- The beacons used are not "intelligent" and require manual entry of water depth.

12.5.4   LONG BASELINE (LBL) SYSTEMS

The LBL system consists of a transducer installed under the vessel and an array of three or more transponders positioned on the seabed. In this calibrated array of transponders, it is recommended to use at least four transponders, which are used for the purpose of redundancy and accuracy. As the transponders are fixed to the seabed and not attached to the moving vessel or a target, the system can operate independently of VRU input. This takes care of many problems associated with vessel motion (Figure 12.11).

The baseline in this case is the distance between the transponders. On board the ship a single transducer communicates with the array of transponders. The LBL system is used to calculate range only, which is the direct distance from the array of transponders. As the depth of transducer is known, this improves the accuracy further.

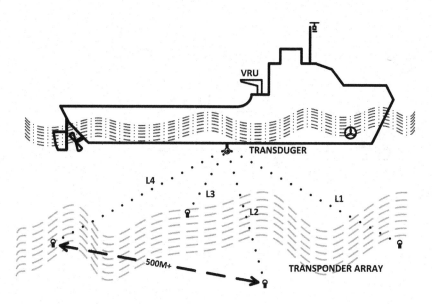

FIGURE 12.11   Long baseline system.

12.6   MECHANICAL PRS – TAUT WIRE POSITION REFERENCE SYSTEM

DP professionals would always agree that the most user-friendly and simple PRS is the taut wire system. The system consists of a weight connected to a wire which is launched on the seabed and any movement of the vessel would result in changes in the angle. This angle is used to find out the distance moved by using geometry and trigonometry. The other end of the wire is wound on the drum of the winch. In between, the wire passes through the inclinometer and the movement of the ship from its present position will result into movement of the inclinometer.

The movement of the ship is measured by the angle of the wire from the perpendicular. By using trigonometry, geometry and calculations the distance moved can be determined. Once this signal is fed into the controller, the controller gives the TAL to activate the thrusters accordingly to move the vessel back to position (Figure 12.12).

It is important to keep the wire in a tight (taut) condition and also important to note that the water depth and the angle of the wire from the perpendicular affect the proper functioning of the taut wire. The taut wire weight, which is also called the mooring weight, is placed on the seabed with the wire connected back to the ship (winch). The moment the weight touches the seabed and weight is taken off the winch, it enters auto tension mode.

This wire passes through the sensor, an inclinometer, which is designed to measure angles and then can be used to calculate the distance moved. If this information (position of the ship) is fed to the controller, an output, the TAL, is generated by the controller, which is used to control and maintain the ship in position.

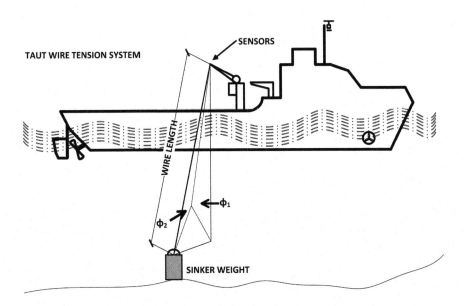

FIGURE 12.12   Taut wire system.

From the diagram, it can be seen that any movement of the vessel would result in the X and Y component. These are then used to calculate the distance moved. Wire length paid out is also measured and used in this process.

The taut wire system is probably the most dependable and error-free PRS among all PRSs currently in use in DP. This is due to the fact that this system does not have any signal interference/noise. As the sensor is moved mechanically to measure the position, there is no chance of electrical interference.

The angle measured/distance moved is sent to the DP controller by an analogue or network signal.

Currently the use of taut wire is done in shallow and medium depth applications only.

At deeper applications, it may not be always feasible to hold the wire taut. Also, the taut wire cannot be used beyond certain angles.

A taut wire is considered a more useful PRS when the vessel is going to spend more time in one position and the water is not deep. Along with the vertical taut wire systems there are also horizontal or surface taut wires. These systems are used to connect telescopic gangways and the position of the floating hotel/accommodation barge can be maintained by the side of the platform/rig.

12.7   FREQUENCY MODULATED CONTINUOUS WAVE (FMCW) RADARS

FMCW or frequency modulated continuous wave radars work on the principle of radio detection and ranging. Most of the time the use of this equipment is restricted to short distances, however these systems have much higher ranges. The transmitter is designed to generate a radio signal at 9.2 GHz. Accordingly, this signal is sent to a powered target by the antenna directed at it. The target sends back a signal to the sensor's receiver and the range is calculated.

The signal from an FMCW radar is transmitted continuously. This differs from pulsed radars. The continuously transmitted frequency is varied across a set bandwidth. The difference between the two frequencies, the transmitted one and the received one, are compared and thus used to calculate the range.

12.7.1 RadaScan

The RadaScan family of sensors by Wärtsilä Guidance Marine, UK (RadaScan and RadaScan View), as the names suggest, are position reference systems which have a 360° field of view. The equipment consists of a spinning radar antenna fitted inside a dome and is controlled at the DP operator station. The three main components of the RadaScan system are, RadaScan main sensor, RadaScan dashboard and RadaScan responder target.

By using the sensor and one or more responder targets on a target vessel, the range, bearing and heading of these targets ay be calculated. These responders are considered as intelligent because they only talk to the interrogation pulse coming in from the interrogator of RadaScan (Figure 12.13).

The RadaScan sensor has an operating range of 1000–10 m and RadaScan View 600–10 m. The RadaScan View sensor also has the unique ability to use radar reflections from the scene to provide situational awareness information to the DP officer. This is particularly useful to identify locations on the responders on the target vessel and for operations performed in low-visibility weather conditions.

12.7.1.1 RadaScan General Care and Maintenance

It is claimed that the equipment is maintenance free and requires no calibration after installation checks are completed. The power, signal cables and connectors are military grade. Responders can be powered by mains supply, primary cell pack (PCP)

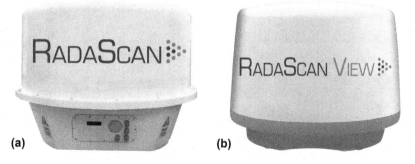

FIGURE 12.13 (a) RadaScan 12.13, (b) RadaScan View (courtesy Wärtsilä Guidance Marine).

battery or rechargeable batteries depending on the locations of the targets and use requirements.

A remote switchable on/off variant is also available for the rechargeable version which will extend the battery life between charges (30 days to many months). There are no maintenance requirements for responders.

12.7.2   RADIUS

RADius is manufactured by Kongsberg and works on the above-stated principle. The main difference between the RADius and RadaScan is that it does not scan 360° and has a 90° working area. The system consists of a computer and its related hardware connected to the interrogator unit fitted suitably on the ship covering the 90° working area. The interrogator unit communicates with the transponder unit installed on the target vessel using frequency modulated continuous wave. Like RadaScan, the RADius also can communicate with multiple targets.

The main interrogator unit is powered by mains and the data cable is used for communication between the interrogator unit and the computer. Depending upon the choice of transponder units fitted on the target vessel a precise range between up to 200 metres can be achieved for good positioning of the vessel. The available options from low power/low gain/high gain transponders may be selected for an operation. The target-acquiring range depends on the type of transponder used and may vary between 400 and 1000 metres.

The frequency range of RADius is between 5.51 and 5.61 GHz. The transmitted power of the main interrogator unit is approximately 1 watt and it is a relatively lighter unit weighing about 7 kg. The transponder unit fitted in the target vessel/rig weighs about 1.5 kg and is battery operated (Figure 12.14).

FIGURE 12.14   Radius components (courtesy Kongsberg).

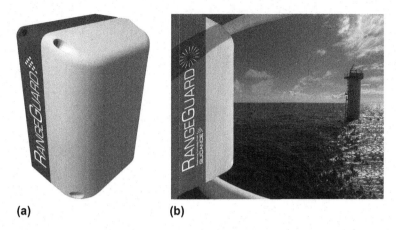

(a)   **(b)**

FIGURE 12.15   (a) RangeGuard, (b) RangeGuard in use (courtesy Wärtsilä Guidance Marine).

12.8   RANGEGUARD

RangeGuard by Wärtsilä Guidance Marine, UK, is the first targetless DP reference sensor based on microwave technology. It is specifically designed for use at monopoles at offshore wind farms. The RangeGuard system consists of the following:

- 2 × RangeGuard Sensor (the main sensor unit).
- RangeGuard Processing Unit (computer).
- RangeGuard Dashboard Software (operator panel) (Figure 12.15).

Each sensor has a fixed field of view of 110 × 11°. There are no moving parts. A single sensor measures the distance to the closest object in its field of view. For offshore wind applications, two sensors are mounted on the vessel such that the radar signals overlap, and simple trigonometry determines the range and bearing to the monopole. The dashboard software is used to monitor/view/control the equipment during DP operations. This signal is then sent to the DP system.

13 Handling of PRS/ PME and Various Tests on PRS/PME

13.1 INTRODUCTION

There are a number of PRS/PME needed according to the class requirements, and also depending on the type of operations, in which a vessel is engaged. Depending on the number of the PRS enabled, the DP system carries out a series of tests to see if the sensors are fit for purpose.

13.2 VARIOUS TESTS ON PRS

To test if the PRS qualifies its measurements to be used, most DP systems carry out a series of tests/checks on all the online PRS/PMEs. The following online tests are performed by a DP system (Figure 13.1):

- Freeze test – commonly known as "live" assessment test.
- Variance test – commonly known as long-term assessment test.
- Prediction test – also referred to as known as short-term assessment test.
- Slow Drift test – known as "assembly assessment test".

Before these tests are performed, all the received position measurements are checked for validity. If the tests fail or the limits set are exceeded, messages are generated for the operators to take action accordingly.

13.2.1 Freeze Test ("Live" Assessment Test)

There may be a situation under certain circumstances that a PRS may repeat the position fix fed to a DP and may do this on a continuous basis. If not detected this may be a serious issue affecting the position-keeping capability of the vessel. Most DP systems are designed to look at this problem very cautiously. If the PRS input is observed to be not varying (fixed) and this happens to be below a predefined limit and for a given time, the DP system rejects this PRS. This helps in identifying and rejecting a frozen input from a defective position reference system.

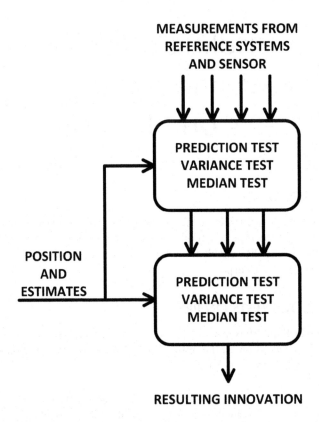

FIGURE 13.1   Tests on position reference sensors.

13.2.2   VARIANCE TEST (LONG-TERM ACCURACY ASSESSMENT)

A DP system checks all the enabled PRSs and uses this information to calculate the variance for each of the position reference sensors. Simply defined, the variance is the squared value of the difference between the measured values of position and the estimated values of the position of the ship. Estimated position is given by the mathematical model.

This test detects if the variance in the measured values from the sensors exceeds the reject limit set for the purpose. The process for setting this limit is very interesting and is based upon the variance of the best reference system. Being continuously calculated, this is based on the variance of the position reference system with the lowest variance value.

This calculated variance value is used for calculating a weightage or weighting factor. The higher the variance the more the value for the weightage will go down. If the set value is exceeded a warning/alarm is generated.

Based upon the test, the DP system assigns the highest emphasis to the PRS which has the highest weightage. Also, a lower weightage PRS would be assigned lower emphasis. The spread of fixes plays an important role in this process.

13.2.3 PREDICTION TEST

Sudden jumps in the PRS may be detected by the prediction test. If the prediction values are more than the set values, the PRS is rejected.

The test will also reject data that show a drift away from the vessel mathematical model's predictions. The limit for the prediction test is a function of the actual measurement accuracy (calculated using variance test).

13.2.4 SLOW DRIFT TESTS

Slow drift tests include the following two tests:

- Divergence test.
- Median test.

These tests are used for detection of possible drift in the PRS in use (Figure 13.2).

13.2.4.1 Divergence Test

A divergence test is initiated when two or more PRSs are enabled in DP. This test detects a systematically increasing difference between the enabled PRS. Thus, the

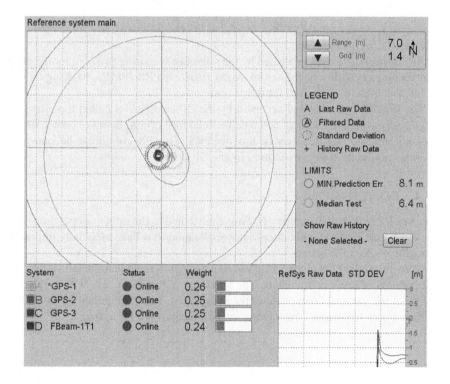

FIGURE 13.2 Median test (courtesy Kongsberg).

process gives an early indication of systematic errors. After this the concerned PRS may be rejected by the prediction test. To sum up, it may be observed that this test only provides a warning to the operator of a systematic degradation and it does not reject the PRS as such.

13.2.4.2 Median Test

A median test, as the name suggests, is a test initiated by DP systems when three or more PRSs are enabled. The median test is used to detect the differences of one PRS from the others. The test makes use of the measurements which are not dependent on the DP system mathematical model. This helps in detecting a low variance and slow drifts.

This test will detect a position reference system whose input data significantly differs from the other position reference sensors online. Most DP systems will have an option for the DP operator to exercise whether to automatically reject the sensor or just use it as a warning. When the set limits are crossed, an audio-visual warning will be issued.

13.3   PROCESSING OF DATA FROM PRS/PME

For a given situation and assignment, a DP vessel may be required to use a number of PRSs. This will also depend upon the equipment class. The DP computer performs a series of tests on each PRS. The purpose of the checks is to verify if the incoming PRSs are fit for the purpose.

All the important data generated by the reference sensors is essential for the position-keeping capability of the vessel. Many different PRSs/PMEs, working on different principles, may be used for the purpose.

The first PRS enabled and accepted by the DP computer is called the reference origin. Any available PRS can be made the reference origin. Being reference origin does not indicate that this may be a better PRS. This PRS of course becomes the internal coordinate system to be used by the DP system.

13.4   CONCLUSION

The way DP systems have improved have contributed to the safety of operations and the vessels. Safe operation in deep water applications has been especially improved using the modern DP system.

Notwithstanding the above, it should also be kept in mind that the DP system is a combination of seven components. The reliability of the DP system will be the reliability of the weakest component in the subsystems. Thus, the reliability of a PRS/PME is a very important aspect in the overall reliability of DP systems.

14 Architecture of a DP System

14.1 INTRODUCTION

The processors used to control a software which operates the DP system are known as the DP controller/process stations or simply controllers. The DP system varies in its level of redundancy and thus the number of items of equipment including the computers installed.

The installation of the DP computers may be undertaken in an appropriate redundancy requirement. It could be a single, dual or triple redundancy. The triple redundancy may be in line with the A/60 bulkhead requirements to meet DP Class 3 guidelines.

Computers need to connect and share the information for the redundancy. Ethernet or a local area network (LAN) may be used to connect two or more computers. For an example, many DP vessels may have a single-computer system. This can be called a simplex DP control.

For better redundancy requirements, a dual system or two-computer system may be used. This may be configured to switch automatically in case of a breakdown of the system online or currently in use.

For more safety requirements, a triple or "triplex" system is used. This triplex system provides an extra element of safety while in operation. Depending on the class requirements, the third redundant system may be installed in the A/60 bulkhead area.

14.2 SYSTEM ARCHITECTURE FOR A DP CLASS 1 SYSTEM

A DP Class 1 system may additionally utilise a completely distributed system concept. This DP system may have a single DP computer or controller connected to a single operator station. These systems may have independent input/output devices, their interfacing units, along with the interfacing units for each environmental and position reference system.

For example, a Kongsberg K-Pos DP-11/12 DP system consists of a DP controller unit and an operator station. Both systems satisfy IMO DP Class 1. The DP computer and the operator station make use of a dual high-speed data network for the communication (Figure 14.1).

The DP-11 systems have a direct/hardwired control of the thrusters using the TAL signal from the controller. A K-Pos-12 system is considered an integrated/networked system and as much cabling is not needed as in the previous case. Other interfaces to the thruster and power systems are also arranged using a dual-data network to other parts of the integrated system.

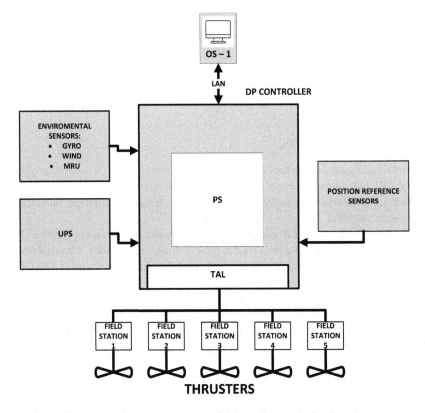

FIGURE 14.1   Architecture of DP-1 system (integrated/networked).

Both systems mentioned above should be able to hold the vessel in position using automatic control of heading and position under defined environmental conditions.

14.3   SYSTEM ARCHITECTURE FOR A DP CLASS 2 SYSTEM

A DP-21 system like DP-11, provides a direct interface to the installed thrusters. The interface also includes the necessary services like power plant, environmental sensors and position-reference systems. The DP-22 system is an integrated system. The connected devices use a dual network.

Both DP-2 systems mentioned above should be able to hold the vessel in position using automatic control of heading and position under defined environmental conditions during and after single point failure has been detected. The failures do not include loss of a compartment.

14.3.1   DUAL REDUNDANT DP SYSTEM

The dual redundancy concept for DP systems is the use of two or more sensors. This type of system also uses a dual computer system, often known as an "online system"

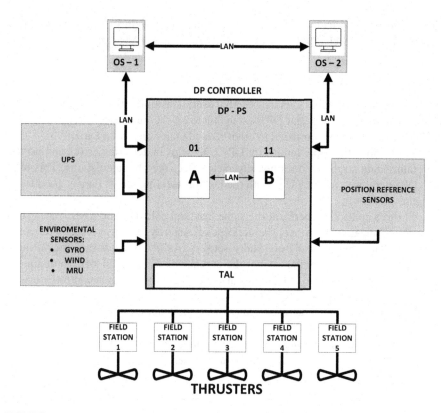

FIGURE 14.2   Architecture of DP-2 system (integrated/networked).

and "hot standby system". This arrangement is used to improve the total availability and reliability of a DP system compared to a single system. Some of the main advantages of redundancy are given below:

- Dual-redundant systems are designed to avoid total system failure when there is a single failure.
- Dual-redundant systems help in detecting a failure, so that corrective action/s may be initiated.
- Fault isolation provides help in not allowing the fault to spread to the healthy system.
- Switchover to hot standby is yet another advantage of a dual-redundant system (Figure 14.2).

14.4   SYSTEM ARCHITECTURE FOR A DP CLASS 3 SYSTEM

Both DP-3 compliant systems mentioned above should be able to hold the vessel in position using manual and automatic control of heading and position under defined environmental conditions during and after a single point failure has been detected.

These failures also include loss of a compartment due to flood or fire. An independent system is expected to be installed in the A/60 bulkhead area.

14.4.1 TRIPLE REDUNDANT DP SYSTEMS

Having better redundancy increases the overall availability and reliability of a DP system. Triple redundant DP systems increase the mean time between failures (MTBF) significantly compared to other redundancies discussed earlier.

For example, the triple redundant DP-31/32 helps in detecting errors and isolates the faulty data so that it does not adversely affect the DP calculations. This helps the DP operator to not be faced with the task of deciding which data is good to use (Figure 14.3).

All three controllers perform the same task and calculate the TAL. When three computers/sensors are in use, the concept of majority voting is used. The voting helps in detecting and isolating faulty sensors. As per the software the voting is performed by all the computers but the computer in command will communicate with the operator stations. Operators can decide and choose which controller is to be made "master".

Given below are the advantages of a triple redundant DP system:

- Voting of sensor input signals to detect faulty/drifting sensor.
- All the three computers use the same data as a basis for calculation of command signals.

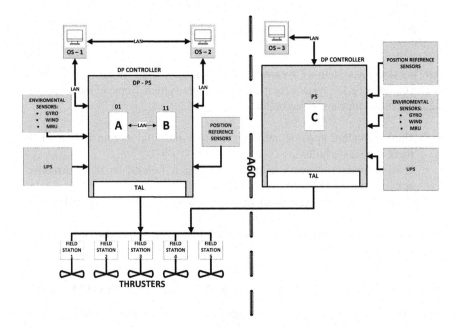

FIGURE 14.3 Architecture of DP-3 system (integrated/networked).

- Triple redundancy brings in software implemented fault tolerance (SIFT) and triple modular redundancy (TMR).
- Voting on command (output) signals is performed differently in the 31 and 32 systems.

In the 31 system, the thruster commands from the three control computers are compared by the "master" computer and after carrying out the median test the command is selected as the final output. In the 32 system, which may be described as a triple redundant networked or integrated system, the voting is performed by using the thruster control panel (TCP) or the field devices (FD) or field stations (FS) in a triple redundant system.

- Can avoid single-point failure.
- Failure detection.
- Fault isolation.
- All computers are "hot".
- Hot repair possible in triple redundant system.

15 Basics of Networking in DP

15.1 INTRODUCTION TO COMMUNICATIONS IN DP SYSTEMS

A DP controller is designed to communicate with related peripheral devices for receiving data input and giving command output. When the DP system contains more than one controller, these controllers must communicate with each other. These communications may be achieved by a network or wiring individual serial data links (Figure 15.1).

15.2 NETWORKING IN DP SYSTEMS

In the simplest way, a network may be defined as a group of two or more computers connected for a purpose. When connected, these computers can exchange information freely, based on the protocol used. There are many types of computer networks and one of the most utilised for the marine use is a versatile local area network (LAN).

15.2.1 NEED FOR NETWORKING IN DP SYSTEMS

Computers are the heart and the soul of the DP system. They make things work as per the installed programme/s. All the inputs from the sensors and position reference sensors are connected to the computer through the appropriate input devices. The mathematical model is stored in the computer and once this information is processed, an output is generated, and this output is used to allocate the required thrust to the available thrusters. The output which controls the thrusters is called the thruster allocation logic (TAL). For the purpose of redundancy, the number of computers used has to be appropriately provided for. When there is more than one computer and they need to be made to work in association with the other computers, networking is required (Figure 15.2).

15.3 BASICS OF NETWORKING

Computer networking utilises protocols to establish communication between two or more computers. A protocol may be defined in different ways in different fields. In computer networking a protocol may be defined as the format that describes how the

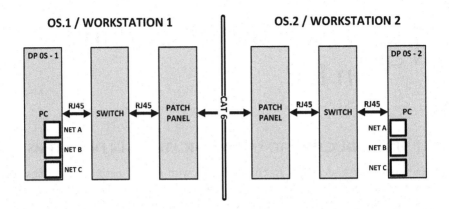

FIGURE 15.1 DP networking between workstations or operator station.

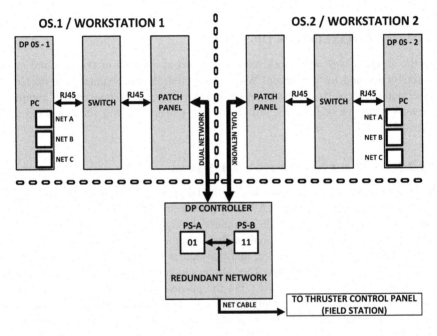

FIGURE 15.2 DP networking between OS and controllers.

messages are exchanged and communicated between sender and receiver. Based on the communication needs between two or more computers, the decision about the protocol is taken. Depending on the requirements, one of the protocols below may be chosen:

- Application layer:
   This is one of the seven layers in open systems interconnection (OSI). The protocols used in this layer mainly focus on the communication

which is process to process. The end-user services can communicate using the application layer. The network applications are supported by this layer. Application layer is often used for protocols like HTTP, SMTP and FTP. HTTP is known to support web services. SMTP has been widely used to support electronic mail, and FTP is used for file transfer.

* Transport layer:

While transporting the application layer messages, client to server and vice versa, the transport layer is used. User datagram protocol commonly known as UDP and transmission control protocol, more often referred to as TCP, are the two transport protocols.

TCP is connection-related service wherein a guaranteed delivery of the message to the destination is assured. Also, this provides flow control of the messages. This is achieved by speed matching of the sender and the receiver. Congestion control is achieved by shortening the long segments of the messages into smaller ones, thereby helping in decongesting the transmission lines. When intended to use a connectionless service, it is the UDP protocol which is utilised. UDP is much faster as there is no control exercised and there is no error correction.

* Network layer:

This protocol may easily be called the most-utilised network in computer applications. This protocol utilises both the TCP and IP, i.e., internet protocol datagram. The network layer protocol defines the IP field datagram and then also defines how the receiving devices and the routers, etc. will deal with these fields.

* Link layer:

This layer is also referred to as layer 2 or the data link layer as this is the second layer out of the seven layers of OSI in the networking of computers. Used in LAN or wide area network (WAN) to transfer data among two nodes of network. As this layer deals with the movement of data this is also called as the protocol layer.

* Physical layer:

The physical layer of OSI is utilised to transfer of data bits with a frame of data, compared to the link layer wherein data from one element of network was transferred to another element of network. It is important to note here that this layer is dependent on the link layer. It also depends upon the media of transmission. The media used could be a modern fibre optic, co-axial cable or a twisted pair copper cable having four pairs of cables to move the data bit by bit.

15.3.1 Networking Used in DP systems

The network used for onboard applications and also for industrial use, LAN compatible with the ethernet IEEE 802.3 standards. Generally, the network is organised as a star-shaped structure with a switch at its centre. This helps in having a network with segments to manage the network traffic in integrated systems, e.g., thruster controls, engine control and power management, etc.

Dual- and triple-redundant stand-alone or integrated DP systems make use of a dual-redundant network consisting of two independent and identical networks generally marked as Net A and B.

The switches, as required, may be mounted at a suitable location inside the cabinet of operator panel or a separate network distribution units (NDUs). Depending upon the type of system, many interconnected NDUs may be used to suit the system specifications.

15.4 COMMON TROUBLES AND THEIR REMEDIES

As of now the network system used on-board are wired systems. This is due to the fact that the environment is not conducive to wireless networks. The disturbances may cause network failure in case of wireless communication.

Wired network cable failure is very common. A network jack may become lose over the period due to heat and vibration and may not make a proper connection. The card receiving the jack may also start giving trouble. The communication between the card and the computer may be lost.

To overcome the above it is recommended to verify that the network cable connection is in good order. LEDs next to the network connection usually indicate it well. If the green LED is working, it means the card is connected and functioning well.

A green LED flashing at a regular interval indicates that the data is being transmitted well.

Some of the devices may have two LEDs. One LED will indicate when the connection is proper, and the other LED will flash when the data transmission is taking place. Please note, if the two lights are orange or red or not there, it indicates trouble for the network and the communication may not be taking place. This invites the attention of the responsible person to look into the trouble and set it right.

15.5 GENERAL TROUBLESHOOTING STEPS
FOR NETWORK PROBLEMS

If you search Google on this topic, you may come across many network troubleshooting flowcharts. The success or failure may depend upon how the logical approach was applied to follow the steps. Very generic and frequently used steps are given below for troubleshooting a network problem:

1. Look for symptoms – note down the symptoms (may be the warnings or alarms will help!!).
2. Localise the scope of the problem – in a chronological way.
3. Was there a change in hardware or software?

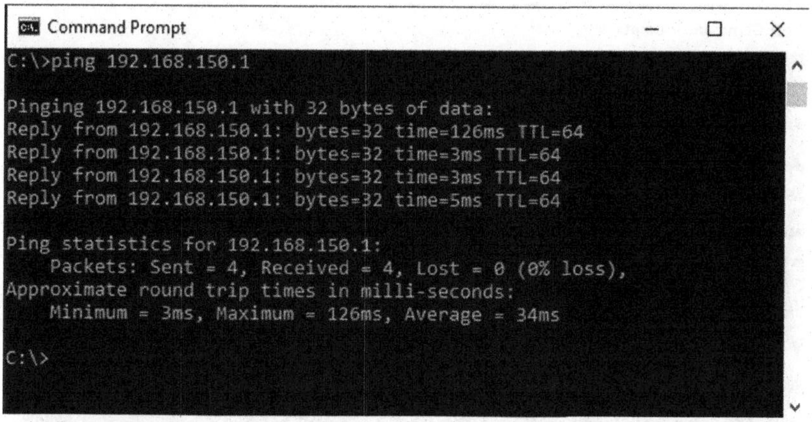

```
█ Command Prompt                                          —    □    ×

C:\>ping 192.168.150.1                                               ^

Pinging 192.168.150.1 with 32 bytes of data:
Reply from 192.168.150.1: bytes=32 time=126ms TTL=64
Reply from 192.168.150.1: bytes=32 time=3ms TTL=64
Reply from 192.168.150.1: bytes=32 time=3ms TTL=64
Reply from 192.168.150.1: bytes=32 time=5ms TTL=64

Ping statistics for 192.168.150.1:
    Packets: Sent = 4, Received = 4, Lost = 0 (0% loss),
Approximate round trip times in milli-seconds:
    Minimum = 3ms, Maximum = 126ms, Average = 34ms

C:\>                                                                 v
```

FIGURE 15.3   Command (CMD) prompt.

4. By combination and permutations, arrive at the most probable cause.
5. Apply the most suitable solution.
6. Check if the applied solution works.
7. Record the solution in logbook/workbook, etc.
8. Make your team aware of the causes and the solutions.

Some of the nice-to-know things about networks and troubleshooting are given below:

15.5.1   PING COMMAND

The ping command can be helpful in ascertaining if the IP-level connectivity is good. If troubleshooting with the help of a ping command, it may be used to send a message which is known as an ICMP echo request to a target computer/ device by using the device's name or IP address. If a response is received back in time it confirms the connectivity and good working condition of all the components in the network between the two devices under test. A ping command may be initiated by using a command prompt commonly known as "cmd.exe". (Figure 15.3).

15.5.2   IPCONFIG COMMAND

The ipconfig command is the short form of Internet Protocol Configuration. This command is also executed using the command prompt. The command can be easily used to find the IP address of the devices connected and the address of the default gateway. Then by using a ping command it can be verified if the two computers/ systems connected by network are functioning properly (Figure 15.4).

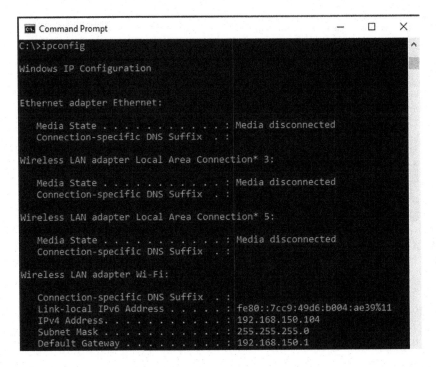

FIGURE 15.4 IPCONFIG command.

16 Types of Signals Used in DP Automation

16.1 INTRODUCTION – DIFFERENT TYPES OF SIGNALS IN MARINE AUTOMATION

Dynamic positioning (DP) vessels are complex ships with lots of signals being made available for the DP controller to process them and take an appropriate decision. The signals may include the environmental sensors' availability and input, position reference system availability and input, thruster signals and various other special sensors' signals. It is important that the DP professionals are fully aware of the signals associated with DP. This will help in fault tracing process.

16.2 SERIAL LINE INTERFACE

A serial port is an interface device in the process of serial communication. The physical interface is achieved using information transfer between two connected devices. The information transfer in a serial communication takes place bit by bit, one after another, serially.

In this case, whether in or out, is one bit at a time not parallel as is in case of a parallel port. Past computer history indicates that most of the time we have used serial interface in cases of modems, terminals and other peripheral devices (Figure 16.1).

The serial line interfaces are mostly used for interfacing the sensors and the position reference sensors with the DP system as described below. Generally, any of the following standards or combinations can be used for interfacing serial line inputs:

- RS232.
- RS422.
- RS485.
- NMEA 0183.

The serial line inputs are galvanically isolated for safety purposes. The input out devices used for the purposes of interfacing are designed to handle the baud rates up to 115,200 bits/second.

16.2.1 What is Serial Line?

The real basics of serial communication need to be understood so that the use of a serial port in a computer can also be understood easily. Understanding serial

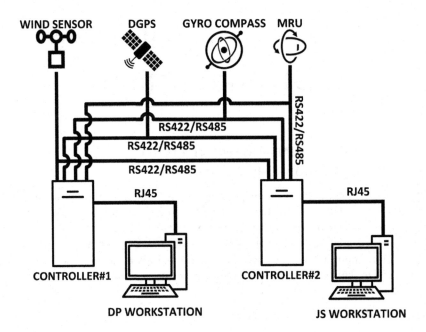

FIGURE 16.1 Serial line interface.

communication will help to check if there is something wrong with the interfaced devices like the sensors, the PRS and various other inputs.

A serial line, also called a serial port, is defined as a device used in serial communication that is used to transfer information between two connected devices. In the past, most computers and laptops had inbuilt serial ports but today only purpose-built computers are supplied with the serial ports. Convertors are utilised for special-purpose use.

Hence, it becomes necessary for engineers to understand about these convertors and other related devices in the interfacing process. The following terms and terminology would be helpful to understand this process:

16.2.2   DATA SPEED – BIT RATE AND BAUD RATE

Serial communication used in the marine environment makes use of bi-state (binary, 0 and 1) signals. Hence, the data rate, calculated in bits per second. This makes the bit rate and baud rate the same. Most sensors and other serial devices permit this baud rate selection in a wide range (Figure 16.2).

To ensure connectivity and smooth data transfer, it is essential that the data rate for the port and the connected device are set the same.

Some devices and equipment may have a facility to select the bit/baud rate automatically.

The rate/speed of transmission also includes information about framing, which are known to us as stop bits, parity, etc. Thus, effectively the rate of data transfer will be less than stated due to these bits utilised for framing. As an example, if your

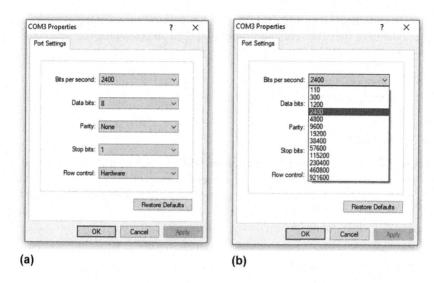

(a) **(b)**

FIGURE 16.2 Bit rate or baud rate settings (a). Bit rate or baud rate settings (b).

setting includes (8-N-1) the effective speed will be 80% of the rate selected, due to the fact that N and 1 which represent "parity" and "stop bit" are also included.

It is the bit/baud rate will decide the rate of transmission. The commonly supported transmission rates in DP automation are between 75 to 115,200 bits per second. The most commonly used settings for standard equipment are 4800 and 9600 bits.

16.2.2.1 Data Bits

How many data bits each character may include is indicated by data bits. The most commonly used data bit is 8 for the modern equipment. A true American Standard Code for Information Interchange (ASCII)may use 7 bits, whereas 5 and 6 data bits are rarely used. With older serial data devices like teleprinters, 5 or 7 bits were quite common.

16.2.2.2 Parity

Parity, odd, even or no parity respectively indicated by 0, 1 or N may be used to detect errors in serial communication. Whenever parity is used in a serial communication, at the end of a signal, an extra bit, a 0 or a 1 is used. By adding a 0 or a 1, the parity bit is either made odd or even.

Most DP systems have provided for interface as per the details given below:

- Data transfer rate between 300 and 115,200 bit/s. This can be configured by software.
- Signal character format.
  Start bit: 1.
  Data bits: 5, 6, 7, 8 (SW configurable).
  Parity: none, odd, even (SW configurable).
  Stop Bit(s): 1, 2 (SW configurable).

- The maximum length of cable used in communication will depend upon the quality of the cable, communication standard and data rate.
  - RS232 – 15 m.
  - RS422 –1200 m.
  - RS485 –1200 m.
  - NMEA 0183 – 1200 m.

16.3   NMEA 0183

National marine electronics association or NMEA format 0183 is a standard for interfacing marine electronic devices. Presently, most of the DP interface makes use of RS 232/422 electrical standard as specified in NMEA 0183, Version 2.0 or above. For more details please refer NMEA 0183 – standard for interfacing marine electronic devices, found at: http://www.nmea.org/.

16.4   POSITION REFERENCE SYSTEM (PRS) INTERFACES

A DP system requires a number of reference sensors to meet the class requirements and also the PRS needed for the task/class of DP. Generally, standard DP systems may have provisions to have the following PRS interface:

- HPR.
- HiPAP.
- Artemis.
- Taut wire system.
- Gangway.
- Fanbeam GPS.
- DGPS.
- DARPS.

HPR/HiPAP/Sonardyne use HPR protocols and also recommended using 400 BCD protocol.

- DGPS/DPS prefer to use NMEA 0183 and a dedicated serial line for DP is recommended.
- Differential absolute relative positioning system (DARPS) also uses NMEA 0183 protocol. DARPS sensors are used by FPSO and shuttle tankers.
- RADius, especially with Kongsberg DP systems, recommended using NMEA 0183 (PSXRAD).

16.5   SIGNALS USED FOR ENVIRONMENTAL SENSOR INTERFACES

- A gyro compass mostly uses NMEA 0183 for the two signals (heading "HDT" and rate of turn "ROT"). This requires a dedicated serial line for DP. As a practice, the heading should be speed and latitude compensated.

- A wind sensor requires a dedicated line to use a NMEA 0183 ("MWV") signal.
- Motion sensors (MRU) use NMEA 0183 ("PSXN").

In addition to the above, a specialised vessel carrying out special tasks using special equipment/sensors/PRS will need to deploy special signals.

A number of interfaces are available which take information from environmental sensors such as wind sensors and gyro compasses. This input is then used by the DP controller. Depending upon the redundancy levels and the DP class required, the number of environmental sensors to be interfaced is decided. Most DP systems provide the following interfaces:

- Gyrocompass.
- MRU/Vertical Reference Sensor/VRS.
- Wind sensor.
- Draught sensor.
- Doppler log.
- Speed signal from DGPS.
- Rate of Turn signal from gyro.
- Water depth.

Serial data – update frequency:

Serial data interfaces are required to be updated after a certain interval. This assures a good interface. For an example, the table below illustrates the minimum and maximum frequency update for Kongsberg KPOS DP system.

16.6   SIGNALS FOR THRUSTER INTERFACES

Most DP systems require the following signals from the thrusters to be connected to the DP controller or process station. The major DP systems and thruster control units make use of two types of signals, digital and analogue.

- Thruster "running".
- Thruster "ready".
- Thruster "enabled" (Figure 16.3).

All the above signals are digital input as the signals travel to the DP controller and are discrete in nature. Some ships/DP systems may have digital output (command) signals in the form of DP control requested for the thruster.

All thrusters are designed to receive a "command" signal from the DP controller. This is called the thruster allocation logic (TAL). On receiving this command, the thruster is expected to act appropriately, and this is indicated back to the DP controller as "thruster feedback" as an input to the controller. The command and feedback signals are analogue signals, whereas the ready, running and enabled signals are digital signals.

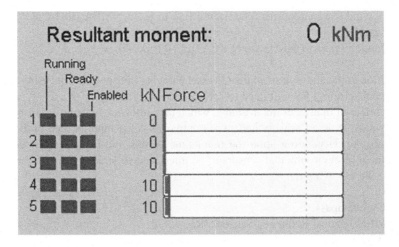

FIGURE 16.3 Thruster running ready and enabled signal display on DP – digital input.

Generally, the digital signals are 24 V. The analogue signals earlier were +/– 10 V but nowadays most of the analogue signals are 4–20 mA. As +/– 10 V signals can get disturbance from the electrical noise/field and can create a false signal, the current signal (4–20 mA) is a good option available to take care of such problems (Figure 16.4).

The health of a thruster can be easily monitored by having a close look at the command and feedback signals. If there is a wide gap between command and feedback signal displays on the appropriate views on the DP operating stations/workstations, it may be understood that either the thrusters are sluggish or have failed.

The type of signals required and used in a ship's propulsion and thrusters will be decided by the type and configuration of the vessel's propulsion system.

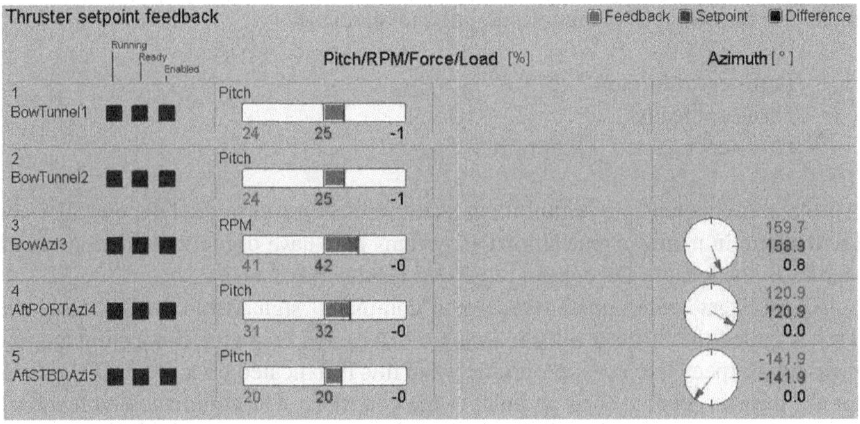

FIGURE 16.4 Thruster set point and feedback signal display on DP.

The following interfaces may be used for a standard DP system:

- Thruster running – digital signal.
- Thruster ready (for DP) – digital signal.
- Thruster enabled (from DP) – digital signal.
- Thruster command (RPM or pitch) – analogue signal.
- Thruster feedback (RPM or pitch) – analogue signal.
- Thruster azimuth command – analogue signal.
- Thruster azimuth feedback – analogue signal.
- Rudder ready – digital signal.
- Rudder azimuth command – analogue signal.
- Rudder azimuth feedback – analogue signal.

Additionally, some vessels may also have the following interfaces:

- Thruster running – digital signal.
- Thruster RPM/pitch reduced (by external system) – digital signal.
- Clutch status – digital signal.
- Thruster RPM (pitch-controlled unit) – analogue signal.
- Rudder in zero – digital signal.
- Azimuth/rudder ready – digital signal.

16.7   SIGNALS FOR POWER PLANT INTERFACES

To manage the power system well, the power management system (PMS) would need interfacing signals from various equipment from the power plant.

Some of the important interfaces are mentioned below:

- Bus-tie breaker.
- Generator power.
- Generator ready.
- Thruster breaker.
- Thruster load feedback (current/power).
- Diesel engine fuel rack.

16.8   OTHER INTERFACES

Various other interfaces are also needed, as required, for special applications:

- Waypoint list.
- Waypoint to DP.
- Waypoint from DP.
- Riser sensor.
- Platform master clock.
- DP alert selector.

- Mooring hawser tension.
- Pipe tension.
- Cable tension.
- Plough tension.
- Water monitor.
- Dredge arm forces.
- Dredge arm positions.
- Line tension.
- Line length.
- Line speed.

17 Consequence Analysis, Capability Plot, Footprints and Motion Prediction

17.1 INTRODUCTION

Consequence analysis (CA) is software-based analytical tool, which is an integral part of the DP programme for a DP Class 2 and DP Class 3 vessels. This programme, when activated, continuously monitors the redundancy of the DP system in the context of power, thruster and buses. CA may be defined as a process which analyses whether in the context of power (generators and bus bars) and thrusters, the vessel can withstand any worst-case failure (WCF) (Figure 17.1).

IMO-MSC 645/1580 makes it compulsory to have an online analysis for DP Class 2 and Class 3 vessels. In simple terms CA monitors, on a continuous basis, if the vessel will hold position and heading in case of WCF. For example, the vessel shown below with three bow thrusters and two azimuth thrusters, is engaged in DP Class 2 operations.

The CA programme will analyse, if for example, the number one bow thruster is lost, what is the consequence as a result? Will the vessel still be capable to hold position and heading?

If the vessel is likely to lose position or heading, the software will generate a warning message "Consequence analysis warning: no. 1 bow thruster critical". The criticality here means that it is critical that bow thruster 1 keeps running. If it fails, the vessel can no longer sustain a WCF or single worst failure, (SWF).

In other words, the alarm indicates that the class is lost and operations must be safely suspended, if engaged in Class 2 or 3 operations. Having lost the class during the operation the DP operator (DPO) may initiate the following:

i. Give amber alert.
ii. Inform master.
iii. Move to a safe location.
iv. Restore class by repair/starting additional resources.

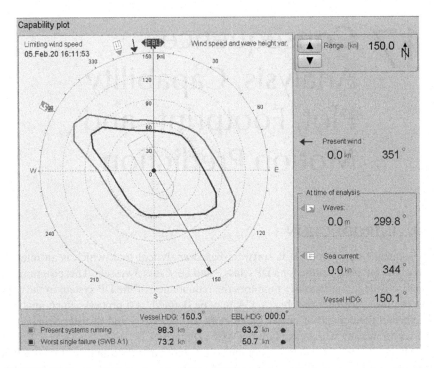

FIGURE 17.1 Capability plot depicting WCF and present system running (courtesy Kongsberg).

To sum up, it may be observed that a CA warning may arise in cases of increases in loads beyond certain limits. This can happen due to:

1. Increase in weather.
2. Failure of thruster, generator or bus.

Also, it is important to note that CA does not monitor anything else apart from power and thrusters. Other components, as listed below, are monitored differently by the DP system:

- Failure of PRS.
- Failure of controller.
- Failure of operator station/workstation.
- Failure of environmental sensors.

17.2 CAPABILITY PLOT

The DP capability plot is the graphical representation of the vessel's position-keeping capability. This function allows the DP officer (DPO) to calculate the maximum weather conditions under which the vessel may be able to continue holding position and heading and hence continue operations. This also helps the DPO to select the optimum heading for a given situation (Figures 17.2 and 17.3).

17.3 FOOTPRINTS

In the context of DP, footprint is the path traced by a vessel around the desired position, while the vessel is operating in auto DP mode. The following diagram gives the graphical plotting of the vessel's movement and represents the footprint of the vessel (Figure 17.4).

The footprint diagram above shows the area marked by different position points at different times and where has been during the period of plotting. It can be well understood that the movement of the vessel around the central position is a function of the following:

1. Vessel's capability.
2. Weather condition.
3. Ship's heading.
4. Draft.
5. Windage area.
6. Direction of current.
7. Swell height and direction.

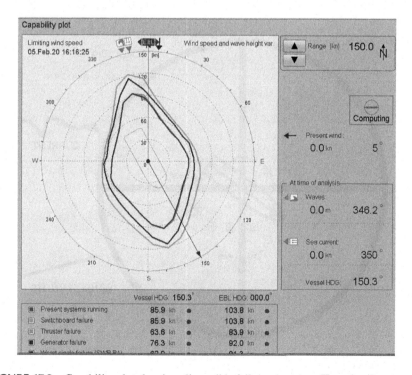

FIGURE 17.2 Capability plot showing all possible failures (courtesy Kongsberg).

DP Capability and Motion Prediction Settings (Changed) [×]

| Main | Run Control | **Thrusters** | Power | Waves |

Thrusters

In-active	Force FB [%]	Mean	STD	Status
☑ BowTunnel1, 1575 [kW]		0	1	■
☐ BowTunnel2, 1575 [kW]		0	1	■
☐ BowAzi3, 1800 [kW]		0	2	■
☑ AftPORTAzi4, 2945 [kW]		3	2	■
☐ AftSTBDAzi5, 2945 [kW]		3	3	■

Simulation possible: ■ [OK] [Cancel] [Apply]

FIGURE 17.3 Settings for capability plot (courtesy Kongsberg).

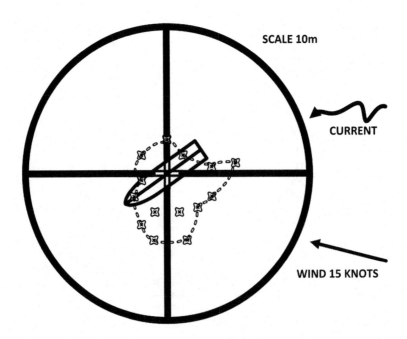

SCALE 10m

CURRENT

WIND 15 KNOTS

FIGURE 17.4 Footprint.

In a nutshell, footprint is a performance graph of the vessel in given environmental conditions.

The footprint can be generated by either of the two following methods:

- By the DPO plotting the vessel's position with respect to the vessel's desired position. It is plotted on a circular graph sheet over a period of approximately two hours. At the end, all marked positions are joined by a smooth curve. All relevant weather details, the location of vessel, that date and ship's head is indicated on the graph.
- Using the functions provided in the DP system and enabling the trace line for the desired periods, a hardcopy of the screen is printed and filed highlighting the prevailing weather conditions.

17.4 MOTION PREDICTION

The motion prediction function calculates how vessels will be moving in a drift-off situation. This is not real, but a simulated situation. This is an additional facility in the DP capability analysis function utilising the same failure configurations as for the capability function. This function helps the operator to calculate the drift-off under the prevailing weather conditions. This also allows the operator to use the expected weather conditions to prepare for it once the results are displayed.

By using his/her experience, the initial position and operating conditions of the vessel, the operator can find the optimal initial conditions for safe operation.

18 DP Trials and Documentation

18.1 INTRODUCTION

The trials on a dynamic positioning (DP) vessel start in the yard itself before the vessel is delivered to the owner. DP proving trials are conducted and then every year trials are repeated. The yearly trials are known as the annual DP failure mode effects analysis (FMEA) trials. The purpose of FMEA/failure modes effects and criticality analysis (FMECA) proving trials is to check the reliability and redundancy. In this process the single point failures are identified and an analysis of their cause/s is carried out in a comprehensive and systematic investigation. Annual DP trials are conducted to assure that the DP vessel continues to have the redundancy for DP Class 2 and DP Class 3 as per the guidelines from the International Marine Contractors Association (IMCA).

DP FMECA trials are conducted by an experienced professional, properly managed and executed to those levels as envisaged in the IMCA guidelines. Whenever a change is made in the DP system components, the FMECA process should be revisited.

18.2 IMCA GUIDANCE ON DP TRIALS

All tests and trials on board are carried out with an aim. The main purpose of FMECA trials is to understand and find worst case failures or the single point failures.

If these identified single point failures are allowed to occur, they result in the loss of the vessel's position and/or heading which affects the ship's capability to maintain position and heading.

During the process, it is important to understand the causes of such failures, what consequences these failures would lead to and what corrective measures need to be initiated. All these must be included in the trials report.

18.3 DP FMEA/FMECA PROVING TRIALS

DP Class 2 and 3 need to have a FMECA trial as mandated by the classification society.

An FMEA report is considered one of the most important technical documents to be carried by the vessel. The need to have a FMEA trial can be traced back to the IMO/MSC 645. The new guidelines IMO/MSC 1580 are now applicable to all vessels. Some classes require that the DP FMEA trials are repeated every five years, whereas others agree that a proving trial every five years may be good for the purpose. It is generally agreed that the five year trial is an extended test. This test is focused on confirming and demonstrating redundancy and that the vessel's DP system continues to function properly. After carrying out the tests the report must be submitted to the classification society for its approval and records.

More documents were subsequently published by IMCA and the following documents may now be referred to for more details about the FMECA trials:

- IMCA M 166 Guidance on FMEA.
- IMCAM178 FMEA Management Guide.
- IMCA M 219 FMEA for a New DP Vessel.

18.4   THE OBJECTIVE OF THE FMEA

The DP FMECA/FMEA trials are aimed at determining the safety, reliability and redundancy of the various components of the DP system.

The main purpose of the FMEA trial is to develop a fault-tolerant system and that even if the worst-case failure happens it does not exceed the specifications. When a fault occurs on the DP system, the vessel should be able to hold position and should have redundancy and be able to correct the fault without affecting the operations.

The main objective of FMEA is to identify various design and process failures even before they are likely to happen. This will help in minimising the risks related to failure.

The risk may be minimised by either proposing design changes or, if they are not workable, then by proposing operationally suitable procedures. FMEA aims to:

- Define and identify the DP equipment, its subsystem and modes of operation.
- Identify the possible modes of failure and their most likely causes.
- Evaluate the effects of failure on the DP system and how they are going to affect the position-keeping capabilities of the vessel.
- Identify the possible measures to reduce or to eliminate the associated risks.
- Plan necessary tests and trials to prove the above.
- Give guidelines to the operators and maintainers on board about the details of the vessel's capabilities. The guidelines must also contain various limitations the DP system may have. This process is aimed at obtaining the optimum performance from the system.

18.5   CONTENTS OF THE FMEA REPORT

The FMEA report submitted on completion of the trials contains the findings of the trials. The findings aim mainly to identifying the failure modes which may be detected during the trials and may have significant effect/s on the performance of the DP system.

The failure modes are graded based upon the failure modes found during the FMEA trials. Not all the failures/failure modes are likely to have the same effects on the system. Some failures may have catastrophic effects and the others may be graded as critical, and some others even may be insignificant.

It is highly recommended that the FMEA should cover all the components of the DP system. There must also be a review of the interfaces and how the subsystem and components share information.

Each test planned should have a purpose, should also have details of configuration of the system. There should be a method defined to carry out the test. Based on the above, when the test is carried out, the results, the real and the expected ones, should be recorded with comments, if any. This must also be signed appropriately.

All failures must be recorded. It is possible that the actual results may not be the same as expected. This is possible due to the fact that the procedure adopted for test may not be proper, the configuration may not be appropriate and sometimes pre-existing faults may also appear as current faults, which the team must be very careful about.

The findings are categorised below:

- Category A findings – These findings are considered potentially serious failure modes. These failures exceed the worst-case failure design intent (WCFDI). The matters which can have safety implications and the issues which may not comply with class requirements also fall under this category.
- Category B findings – These are those findings which relate to failure modes which do not exceed WCFDI. These failures also do not concern to the safety issues. This category of failures is still considered important enough, with the aim of making the DP system safe by adequate redundancy.
- Category C findings – These findings are aimed at improving the system and are considered as suggestions for improvement.

18.6 USES OF FMEA

As a primary tool, the FMEA may be used to assess the capability of the vessel to perform its designed functions safely. To use it for this purpose, it is mandatory that the FMEA has been reviewed and updated as per standard practices. Any update must include the changes and modification/s carried out on the ship's DP system through the entire lifecycle.

The requirements for allotting DP classes notations by the classification societies are well established. In this way the classes are following the IMO MSC guidelines by implementing DP class notations.

The class may require the vessel to maintain certain documentation in this process. All these are specified in the scope and checked accordingly during the audit process. The FMEA report is one of the most important such documents before the class notation for DP 2 or DP 3 is awarded.

The FMEA report is viewed as an important document to assess the vessel for charter or purchase. The FMEA report is a checklist and used during the ship's pre-charter process. It is referred to while the purchase process of such vessels is in progress.

FMEA helps the onboard staff, DPOs and engineers to know the corrective measures to be initiated after failure/s have been detected or have occurred. For example, when a generator fails, or redundancy is lost and the watchkeeper on the bridge or in the engine room needs to initiate a series of actions. Hence, being familiar with the FMEA pays rich dividends for the onboard staff.

FMEA reports also form an important part of the risk assessment process.

An FMEA report can also be used as tool to understand how the systems work and also to identify their failure modes. This can be a good source of training for the ship's staff.

The outcomes of the FMEA trials should be included in the operational procedures and manuals as these have an effect on the vessel's operational capabilities.

18.7 THIRD PARTY AUDIT

Every DP-enabled vessel is required to undertake DP proving trials, in addition to commissioning and testing. The DP system is proven in all normal modes of operation for which it has been designed for. This will prove that the modes, as designed, are working well. Having proved that all normal modes are functioning correctly, next the simulated or real tests are conducted to test the various failure modes. A test of performance in normal working mode and failure mode is also carried out. Such tests/trials are generally carried out or witnessed by some competent third party and thus also called third party audit.

18.8 ANNUAL PROVING TRIALS

The market practice is that all DP2 and DP3 vessels are required to undergo annual DP proving trials. Such trials can be carried out by a suitably qualified auditor. The earlier market practices were that a master or a chief engineer with experience on DP vessels or a suitable person decided by the company may carry out such audits.

As per the new guidelines (IMO MSC.1/Circ. 1580) proving must be undertaken annually. The window period is six months. It can be undertaken within three months before or after the date of the initial survey of the vessel.

The IMCA document IMCA M 166 now defines the authorised personnel to carry out the FMEA trials. The personnel should be adequately qualified and experienced. IMCA conducts an accreditation scheme exam for DP Practitioner and Company DP Authority for this purpose.

18.9 MOBILISATION TRIALS

While being deployed for a new job the vessel's DP system is required to be tested so that it functions as designed and expected. When carried out the tests should demonstrate that the DP system under test is fit for the purpose for general DP operations.

These trials are focusing more on the capability and status of the vessel's DP equipment between annual trials. Market practices are that such trials may be witnessed before accepting the vessel for safe operations by a new client. The mobilisation trials are designed to test that all equipment are working as per the designed capacity.

In short, it may be observed that mobilisation trials are a series of tests to verify the DP system. These trials may be combined with other trials or may be conducted after annual trials. This, at times, may result in some duplication but it does demonstrate the vessel's continuing DP capability.

Although there are no standard guidelines yet, mobilisation trials can be carried out prior to passage to the work site or prior to entering an installation's 500 m zone. Some clients do require a separate set of DP trials, it is expected that the vessel still conducts the mobilisation trials.

The records of the trials are expected to be maintained on board for two years and should be retained on board for two years and then archived for further four years.

Printing out and keeping records of alarm page/parameter recording are a requirement. Most oil majors/charterers may require keeping a record of a printout of alarms and parameter recording on board DP vessels. The records may be the same as the company's document control procedures.

18.10   DP VENDOR REPORTS

There is a practice wherein the record of all DP-related vendor visits to the vessel are recorded. Most of the time the vendor's reports are filed as given and in some cases the crew on board may write a report based on the vendor's report and file it.

18.11   DP CORRESPONDENCE/COMMUNICATION

This may include fax, internal and external memorandums, emails and official letters. The market practice is to maintain a record of all DP-related correspondence to and from the vessel.

18.12   DP FAULT LOG

Many DP-enabled vessels maintain a record of minor faults or problems with DP equipment. This may be done by Electro Technical Officer (ETO)/DP Operator (DPO) as per company guidelines. When a fault occurs the ETO, the chief engineer and master should be informed immediately, and all details may be recorded in this log. This log serves two purposes., one is to record the type and frequency of the fault occurrence and the other is to make all DPOs aware of faults within the system. When the fault has been attended to and corrected by the ETO or vendor technician this must be noted alongside the fault report.

18.13   DP INCIDENT LOG

All DP-related incidents are logged to keep a record. This is to keep an account of the incidents which occur during operations and are recorded as per the safety management system (SMS) procedures of the company.

18.14   OPERATIONAL ISSUES

DP operations are mostly carried out in very challenging locations and weather conditions. To ensure safe operations all DP professionals must be aware of the following:

- Worst-case failure (WCF) – This means the identified single fault in the DP system which may result in a serious effect on DP system capability. This is identified and tested during the FMEA and the FMEA proving trials. All DP professionals are expected to be familiar with the WCF of their vessel.
- DP drills and emergency response drills (DPD and ERD) – It is important that the drills are planned and carried out as close as possible to a working scenario of the vessel. This is explained in the activity specific operational guidelines. This is also important that the DP professionals are aware of the appropriate actions to be initiated when the operational parameters are exceeded. They should all have easy access to the FMEA/FMECA trials for reference. Vessel DP drill scenarios may also be considered based on the reported incidents from IMCA.
- Critical activity mode (CAM) or critical activity mode operation (CAMO) – This may define the configuration for the DP system and associated equipment, which is most fault tolerant under the given situations. In practice vessels may have only one CAM and one task appropriate mode (TAM). All critical activities of the vessel must be carried out in CAM mode only as this assures that the WCF of the vessel is not exceeded and all operations are safely carried out. If given situations do permit for a lesser fault-tolerant standard, it may be advisable to use TAM.
- Task appropriate mode (TAM) – TAM may be defined as a risk-based mode of operation in which the DP vessel may be permitted to use a set-up knowing that the potential failure may exceed the vessel's identified WCF.
- Activity specific operating guidelines (ASOGs) – ASOGs are designed to provide guidelines on how to manage the operational, environmental and equipment performance of the vessel. ASOGs are designed for the identified location and given activity and levels of risks identified. A DP vessel may have many ASOGs for the given location and operating condition and different levels of risks involved.

Depending upon the location of the vessel, if near a well the same may be referred to as well specific operating guidelines (WSOG). For a given field, it may be called field specific pperating guidelines (FSOG) and for a particular location it is called location specific operating guidelines (LSOG). It is important to note here that the above mentioned, CAM, TAM and ASOG are to be based on the good knowledge of the following:

- Ship and the DP system.
- FMEA/FMECA trials of the ship.
- The mission or the job to be undertaken.

- Location of the ship.
- Risk assessment and risk management strategies.

ASOG is generally prepared in tabular format and the four status categories are considered:

1. Green or normal operations status.
2. Blue or advisory status.
3. Yellow or degraded status.
4. Red or severely degraded status.

The green status indicates that the DP system is operating in agreed CAM or TAM mode and there is nothing to worry about. If the vessel is approaching its operational, environmental or equipment performance limit envelopes, it is an advisory situation indicated by "BLUE". If the operational/equipment conditions which may lead to loss of redundancy will create a situation and this is indicated by "YELLOW". When the ships redundancy is severely degraded, and ship is not considered safe to operate/ continue with the activity, this is the "RED" status.

19 Circuit Tracing and Fault Finding on DP Equipment

19.1 INTRODUCTION

An electrical circuit diagram is designed to give an understanding of the basic functioning of the electrical circuit. This may be represented graphically or may be in a schematic form.

An electrical diagram shown pictorially, or a hydraulic diagram may contain a circuit with simplified images of components, whereas a schematic diagram may indicate how the interconnections between various components and subcomponents are made.

In an electrical circuit diagram, which is different than a block diagram, the location of various components may differ from the physical arrangements. An electrical diagram represents real electrical connections, by which the physical arrangements and wiring connections will be easy to locate and trace. This may also be referred to as wiring diagram (Figure 19.1).

A circuit diagram depicts a real electrical connection, so it is different from the block diagram. Maintenance and technical manuals may have other drawings which may indicate how the arrangements of various components are made physically. These are called wiring diagrams as well.

As shown in the single line diagrams can be very helpful to locate and check connectivity. This indicates the cable number, how two devices/subsystems are connected and finally feeding the DP controller.

From here we can go to the next step, i.e., "cable list", which indicates what information the cable carries from where and to where. To find out more about this the next step can be referred to as the "cable details". Cable details will give an indication of various signals carried by the cable and connection. To learn more details about the connections and types of signals refer the last step, i.e., input/output specification". If followed in sequence and carefully this can be a very good tool for fault tracing.

It is important to understand how these components are connected using wires and connections. It is also important to know the arrangements of the wires and connections. These are generally known as a wiring diagram. Very often, electrical circuit diagrams are used in designing and manufacturing printed circuit boards (PCB).

For example, a single line diagram for a power system is the representation of the complete power generation and distribution of the vessel. This single line diagram shows the main connections and arrangements of various parts of the power generator, how the generator is connected to the main switchboard/bus. More often than not, the details like rating, voltage, resistance and other important characteristics are also shown.

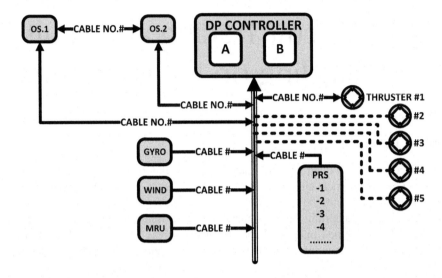

FIGURE 19.1   Single-line diagram – sample.

Referring to a DP equipment technical/installation manual, knowing the layout and wiring diagram of DP system is considered an important approach by any engineer to look into and investigate a problem/fault.

DP professionals working on DP-related equipment, must master the skill of reading schematic and wiring diagrams.

This will pay rich dividends by assuring that risks of shock or damage are significantly reduced.

If the electric circuits within the equipment are known, maintenance and repair work can be carried out more safely and with confidence. Even if something cannot be repaired, the result obtained from this process may be shared with the equipment manufacturer so that the assigned technician is made aware and can come prepared.

19.2   HOW TO READ SINGLE-LINE DIAGRAMS/SCHEMATICS

Single-line diagrams/schematics are the outlines which provide guidance on how the electrical/electronic equipment is functioning, assembled and serviced.

A single-line diagram/schematic helps the operator and the service technician to understand the circuit functioning. This will help to familiarise them with the equipment.

Below are common tips to read schematics.

1. Like normal English text we should read also schematics from left to right and top to bottom. It is important to understand the path that the signal travels and the same follows for tracing the schematics.
2. Become familiar with the schematic standard sign and symbols. It pays rich dividends to know this language. This makes the reading of schematics very easy. With minor changes here and there, the symbols are standard.

Understanding ground or the return path is very important for understanding the path of the signal.

Ground or earthing may be represented by a downwards pointing triangle or sometimes by a few parallel lines bigger on top and smaller at the bottom.

It should be clear that the ground is a common reference and may not always be an actual ground. A line may represent a cable or a wire as the case may be. These line/s may cross over each other but it may not indicate that they are connected.

Understanding Ohm's Law can always be a starting point for knowing electrical circuits. It is important to know and remember that the voltage drop across a resistor is equal to the current flowing through the circuit multiplied by the resistor ($V = IR$). How capacitors are represented, how resistors and other such components are used to condition the signals, will make it easy to understand the schematics.

3. How the interrelation of each component is established in the complete circuit/device will make it easier to know the circuit well.
4. Establish and know the various tasks performed by the circuit. In this process, the OEM's data sheet for the equipment/subsystem/circuit will be helpful.
5. Schematic reading is not complete without knowing the function of each component. This will eventually lead to knowing the function of the full circuit.

19.3 FAULT FINDING PROCEDURES

Most OEM's manuals will give an indication of how to proceed with fault finding. All warnings and alarms give enough indication and that should be taken as the basic guideline. Alarms and warnings provided by the system are meant to draw attention of the DP officer/DP engineer and hence the warning and alarms must be monitored carefully.

There are guidelines available as promulgated by International Marine Contractors Association (IMCA) and Marine Technology Society (MTS) -DP which need to be followed by the DP professionals to ensure that no untoward/unintended actions result in a breakdown of a system/subsystem/sensors while the DP operations are on. In case something like this is detected, it is safe to move the vessel out of 500-metre zone and a complete investigation to be undertaken, the fault rectified and re-entry to the work area are to be made as per guidelines available.

It is an industry safe practice to ensure that no operations may restart before a system is satisfactorily repaired, tested and recorded to ensure safe operations.

The onboard crew and responsible person/s may seek further guidance from a 24-hour helpline from the OEM or the office ashore in case some difficulties are experienced.

It is important to remember that IMCA always emphasises the importance of ensuring that no work on DP systems is carried out during the operations. If need arises, it is always recommended to follow ASOG.

The DP operations team should be trained as team. Most ships have the standing orders, and this must be followed to ensure that no work is carried out during operations. When there is a need, in unavoidable circumstances, follow the procedures and permit to work (PTW)/ electrical permit to work (EPTW) must be initiated.

19.4   SOFTWARE FAULTS AND RELATED ISSUES

More often it is observed that the vendor/OEM for software may not like to share the records of installing various versions of software and at the same time the records thereof may not be updated. The DP professionals on board should develop a good habit of keeping a record of such things. It is a practice that every after a major upgrade of software, FMECA trials must be completed ensuring safe operations. Many good companies follow a practice of management of change (MoC) process. This is helpful to understand the benefit/s of the new versions of new software.

Many experts agree to the fact that the MoC may not be able to address all issues with software like bugs, etc., but there seems to be general agreement that it is good to have a suitable MoC.

20 Roles and Responsibilities

20.1 INTRODUCTION

For safe and efficient operations, the roles of everyone involved in DP operations are well defined. The overall responsibility is with the master. The SDPO/DPO on the bridge and the engineer on the watch in the control room function as a team. The chief engineer has overall responsibility for ensuring the required machinery in the engine room, on the deck and in the control stations are functioning to the best of their capabilities. The electrical technical officer (ETO) plays a pivotal role in achieving this.

All the personnel responsible for DP operations will be allocated with particular duties and specific responsibilities during operations.

The general duties and responsibilities are given below, however, they may vary depending upon the vessel's type, design, hierarchy and the client's requirements.

20.2 CAPTAIN/MASTER

The captain/master on board a DP vessel may or may not be required to keep a normal navigation/DP watch. As is the practice in the market, the master is available on call at all times as needed by the watchkeeper for expert advice. The master of a DP vessel is entrusted with lots of duties and responsibilities, a few of them are listed below:

- Conducting annual trials.
- Leading and controlling DP drills.
- Onboard training for new DP operators.
- Assessing the skill level of new DP operators.
- Assessing the skills of the existing DP personnel.
- Hazard identification (HAZID)/risk assessment.
- DP incident/accident investigation.
- DP emergency procedures.

20.3 SENIOR DP OPERATOR (SDPO)/ DP OPERATOR (DPO)

- Keeping DP watch.
- Ensure safe operations.
- Maintaining log.
- Monitoring DP system and position of the ship.

20.4   CHIEF ENGINEER

The chief engineer (CE) is an experienced marine engineer, fully trained, competent and qualified.

He/she is overall in charge of the machinery on board a DP vessel. He/she is responsible for the following in general:

- In overall charge of the engine room (ER).
- Power generation.
- Mechanical, electrical and electronic departments.
- The CE also ensures that engineering personnel working under him/her are trained to the required standards and that they are all qualified and competent to carry out the jobs assigned.
- The CE is also responsible for preparing and maintaining the procedures for general operations and emergency situations related to DP and other engine room activities.

20.5   ENGINEER OF THE WATCH

As mentioned earlier, although the vessel may be Unattended Machinery Space (UMS) class, yet during DP operations the engine room is always staffed by a watch-keeping engineer. The main responsibilities of the watchkeeping engineer include, but are not restricted to, the following:

- Ensuring normal operation of the vessel's switchboards, maintaining appropriate functioning of bus tie breakers.
- Diesel engines and power generation and distribution systems.
- Thruster start-stop and parameter monitoring, auxiliary systems, such as fuel, compressed air and cooling water systems.
- General ER watch keeping duties while the vessel is on DP. This may include emergency situations like blackout, recovery from a blackout, response to various failures.
- Recovery from failures of online machinery and safe changeover to standby as required.
- The watchkeeping engineers must understand the consequences of the failures above.
- Ensuring the requisite redundancy levels are maintained.
- Understand and ensure the operational requirements of the vessels are met and maintained.

20.6   ELECTRICAL TECHNICAL OFFICER (ETO)

The electronics technician, electrical officer, ETO or any other designation the company may assign for the person responsible for the safe working and maintenance of the electronic parts of the DP system, related subsystems and equipment and their components.

The ETO on board a DP vessel, is likely to report to the CE for all technical matters and responsibilities. For the matters related to DP the ETO may directly report to the captain as some vessels may have such arrangements. The responsibilities may vary from vessel to vessel and of course the equipment installed on board and the job assigned to the vessel.

An ETO, a fully trained, qualified, experienced and competent electronic/electrical engineer, may have the following general responsibilities:

- Upkeep and maintenance of a log for the DP system maintenance. This includes all the maintenance carried out on the DP system.
- To maintain the DP control system and associated systems including sensors and position reference sensors in good working condition
- To carry out routine checks and maintenance on DP-related equipment including the following but not restricted to:
  - DP control system interfaces.
  - DP computer functions.
  - Tests and fault finding on DP system or its components.
  - Thruster system and related power unit including variable frequency drives.
  - Sensors (environmental sensors and position reference sensors).
  - Power generation, distribution and UPS systems.
  - DP system hardware and the software.
  - Good knowledge in referencing equipment supplier/vendors manuals.
  - Good understanding and knowledge of vessel FMEA.
  - To carry out tests and maintenance/repairs as needed, by replacing spares as per recommended practice by the vendors.
  - Communicate and liaise with the bridge and the ER team.
  - Completing check lists and paperwork as assigned.

20.7   DP TRAINING FOR ENGINEERS AND ETOS

The Nautical Institute constituted a work group to formulate the guidelines for DP training for the engineers and ETOs on board a DP-enabled vessel. This was aimed at ensuring that every engineer and ETO on board should be able to contribute to safe DP operations, carry out the required maintenance on the DP-related equipment and attend to logical fault finding. This is certainly not easy unless the engineers concerned have undergone a structured training programme to understand the DP system well, so that the logical steps for fault tracing and fault finding may be initiated when needed. Based on the above the Nautical Institute published the guidelines to be followed by all the accredited/recognised DP training centres. The latest series of guidelines was released in January 2020. (Refer: The Nautical Institute – Certification and Accreditation Standard Vol 1.-DPSTTC-V1-02/01/2020 -ANNEX I – DP KNOWLEDGE FOR TECHNICAL STAFF COURSE.)

Reference may also be made to the following documents for more detailed understanding of the training related issues:

- IMO MSC 645: International Maritime Organization, Guidelines for Vessels with Dynamic Positioning Systems, Maritime Safety Committee (MSC) Circular 645.
- IMO MSC 1580: Guidelines for Vessels and Units with Dynamic Positioning Systems, Maritime Safety Committee (MSC) Circular 1580.
- IMCA C 002: Guidance on Competence Assurance and Assessment: Marine Division.
- IMCA C 007: Guidance on Assessor Training.
- IMCA M103: Guidelines for Design and Operation of DP Vessels.
- IMCA M117: The Training and Experience of Key DP Personnel.

Glossary and Abbreviations Used in Dynamic Positioning

Every subject area may have a commonly used abbreviation and glossary. The field of dynamic positioning also makes use of some common abbreviations and terms. For the benefit of readers, the following abbreviations may be of good use.

ABB   Asea Brown Bowey. Producer of various equipment for production and power management for the offshore industry.

ABS   American Bureau of Shipping.

ADP   Short for Albatross DP, also short for Alsom DP.

ADS   Atmospheric Diving Suit, enables divers to access depths greater than 450 m.

Alstom   DP manufacturer, formerly Cegelec/Converteam now GE Converteam.

ADOC   The association of offshore diving contractors, merged in 1995 with DPVOA, now known as IMCA.

APL   A Norwegian company that produces various offshore loading systems, e.g., SAL, STL, OLS.

ARAP   Absolute and Relative Acoustic Position reference system.

ARTEMIS A microwave position reference system using one Fix and one Mobile transceiver giving range and bearing to the target.

AS   Auto sail. Could be a DP mode.

ASCII   American Standard Code for Information Interchange.

ATA   Automatic thruster assistance.

Auto Start Part of a vessel's power management system, where standby generator starts automatically when more power is needed.

AUT   DNV class notation for vessels with DP Class 1 equipment.

AUTR   DNV class notation for vessels with DP Class 2 equipment.

AUTRO   DNV class notation for vessels with DP Class 3 equipment.

Blom PMS Position Monitoring System. It monitors raw data from positioning reference systems and sensors input to the DP, i.e., unfiltered data.

CA/Code The Course Acquisition code used by GPS.

CA   Consequence analysis.

CEGELEC A DP system.

COG   Centre of gravity.

COR   Centre of rotation on a DP vessel.

COT   Centre of turret.

CRT   Cathode ray tube.

D Surge Damping of surge.

D Sway   Damping of sway.

D Yaw	Damping of yaw.
DARPS	Differential Absolute and Relative Position Reference system.
DGPS	Differential Global Positioning system.
DNV	The Norwegian Class Society.
DoD	Department of Defense (USA).
DP (ΛΛΛ)	Lloyds class notation for a Class 3 DP vessel.
DP (AA)	Lloyds class notation for a Class 2 DP vessel.
DP (AM)	Lloyds class notation for a Class 1 DP vessel.
DP (CM)	Lloyds class notation for a DP vessel of limited capability.
DP/DYNPOS	Dynamic positioning.
DPO	Dynamic positioning operator.
DPVOA	DP Vessels Owners Association. Merged with AODC in 1995 to form
IMCA,	International Marine Contractors Association.
DSV	Dive support vessel.
Duplex DP	A DP system having two computers to provide redundancy.
EGNOS	European Geostationary Navigation Overlay System. A system that improves the accuracy of a GPS position to 5 m or better. It is free to use.
ERS	Environmental reference system, i.e., wind sensor.
ERN	Environmental regularity numbers (DNV). A capability indicator.
ESD	Emergency shutdown system.
Fanbeam	A laser type of position reference system. Giving distance and bearing to target(s).
Flotel	Floating accommodation vessel.
FMEA	Failure mode and effect analysis. A failure mode test of the DP system.
FOC	Full operational capability of the GPS system.
FPSO	Floating production, storage and offtake. A production vessel with storage and offloading capability.
FSU	Floating storage unit. A vessel with storage and offloading capability.
FSVAD	Flag state verification and acceptance document.
FZP	Field zero Point. The geographical position of the centre of an oil field.
Gallileo	The European version of GPS.
Glonass	The Russian counterpart of the GPS system.
GPS	Global Positioning System (USA).
Habitat	A dry compartment on the seabed at the worksite used by divers.
HDOP	Horizontal dilution of position. A measurement of accuracy in a GPS position.
HELIOX	Helium-oxygen gas mixture used by divers for breathing at depths greater than 50 m.
HiPAP	High precision acoustic position reference system.

HPR	Hydro acoustic position reference system.
HSE	Health and Safety Executive. The statutory body responsible for safety in the UK sector of the North Sea.
ICMP	Internet Control Message Protocol.
IMCA	International Marine Contractors Association.
IMO	The International Maritime Organization.
IOC	Interim operational capability of the GPS system.
IMS	Information management station.
ITCS	Integrated Tthruster Ccontrol system (ABB).
LBL	Long baseline HPR.
Mercator	A navigational chart projection based on Lat-Long coordinates.
MOU	Mobile offshore unit.
MRU	Motion reference unit.
MSA	The UK Maritime Safety Agency.
MTC	Manual thruster control.
Nautronix	A U.S. maritime technology company, also DP producer.
NI	The Nautical Institute.
NMD	The Norwegian Maritime Directorate.
NPD	The Norwegian Petroleum Directorate.
OLS	Offshore loading system. A sub-water offshore loading buoy system located at the Statfjord and Hibernia fields.
OLT	Offshore loading tower. A surface loading buoy for offshore loading by shuttle tankers.
P-M	Pierson-Moscowitch wave spectrum model.
PM/Posmoor	Position mooring systems PMC/PMCON Posmoor Controller.
PMCQA	Posmoor consequence analyser.
PMSIM	Posmoor simulation.
POSCON	Position control system. Usually a joystick system with auto heading mode.
PRC	Pseudo range corrections within the GPS system.
PRS	Position reference system.
RADius	A newly developed position reference system, based on radar principles giving range and bearing to one or multiple targets.
RC	Rotation centre.
Responder	A type of transponder where the interrogation signal is sent via cables instead of through the water.
RO (R-O)	Reference origin.
ROV	Remote operated vehicle, usually an unmanned mini submersible.
SA	Selective availability.
SAL	Single anchor loading. A sub surface loading system.
SBS	Short baseline (HPR).
SDP	Simrad Dynamic Positioning system (Kongsberg Maritime).
Simplex	Reference to a non-redundant DP system.
Sonardyne	A UK manufacturer of underwater acoustics.

SPS	Standard Positioning Service, GPS.
SPM	Single point mooring, a loading tower where a shuttle tanker connects at the bow with a single hawser and a loading hose.
Starfix	A commercially operated position fixing system covering US waters from satellites in geo-stationary positions
Stinger	A heavy gantry deployed at the stern of a pipe lay vessel, supporting the pipe as it is laid. Used in the S-lay method.
STL	Submerged turret loading. A loading system where a submerged buoy is hoisted by the tanker and mated into a mating cone in the forepart of the vessel. When the turret is locked the vessel is securely moored.
Surge	A vessel's movement in the fore-aft direction.
Sway	A vessel's movement in the transverse direction (starboard/port).
SSBL	Super short baseline (HPR).
TCP/IP	Transmission control protocol/internet protocol.
TW/Taut Wire	A position reference system consisting of a crane with sensor detecting the length and angle of tensioned wire leading to a clump weight on the seabed.
TMR	Triple modular voting, the voting concept of redundancy.
TP	Turn point, defined within an auto track function.
Tp	Transponder used as a part of the HPR system.
Transducer	The acoustic probe on the vessel's bottom. Used as a part of the HPR system.
Trencher	A seabed crawler vehicle intended to trench and bury previously laid pipelines.
Triplex	A DP system having three computers providing redundancy through voting logic.
USP	Uninterruptible power supply, battery backup of power to the DP system.
UTM	Universal transverse Mercator projection and coordinate system.
USBL	Ultra short baseline system (HPR).
Voting	A system using triplicate systems and a "two out of three" vote on all critical values.
VRS	Vertical reference sensor. A devise giving accurate values for pitch and roll movements on a vessel.
VRU	Vertical reference Unit. Similar to VRS.
WAAS	Wide area augmentation system. The American equivalent of the EGNOS. A system that improves the accuracy of a GPS position to 5 m or better. It is free to use.
WCF	Worst-case failure.
WGS84	The World Geodetic Spheroid upon which the GPS system is based.

Bibliography

American Bureau of Shipping. ABS Rules for Building and Classing Steel Vessels. 2001. Part 4. Chapter 3. Section 5. 15 Dynamic Positioning Systems. https://ww2.eagle.org/en/rules-and-resources/rules-and-guides.html.

DNV. July 2001. Rules for the Classification of Steel Ships – Part 6, Chapter 26. Dynamic Positioning Systems. http://rules.dnvgl.com/docs/pdf/dnv/rulesship/2016-07/ts626.pdf.

Guidance on Failure Modes and Effects Analysis (FMEA). https://www.imca-int.com/publications/179/guidance-on-failure-modes-and-effects-analysis-fmea/.

Guidance on position reference systems and sensors for DP operations https://www.imca-int.com/publications/463/guidance-on-position-reference-systems-and-sensors-for-dp-operations/

IEC Standard, IEC 60812. 2006. Analysis Techniques for System Reliability – Procedure for Failure Mode and Effects Analysis (FMEA). https://webstore.iec.ch/preview/info_iec60812%7Bed2.0%7Den_d.pdf.

IMO MSC Circular 645. Guidelines for Vessels with Dynamic Positioning Systems. https://www.imca-int.com/publications/76/guidelines-for-vessels-with-dynamic-positioning-systems-msc-circular-645/.

IMO. December 2002. MSC Resolution 36(63) Annex 4 – Procedures for Failure Mode and Effects Analysis (HSC Code). http://www.imo.org/en/KnowledgeCentre/IndexofIMOResolutions/Maritime-Safety-Committee-(MSC)/Documents/MSC.97(73).pdf.

Lloyds Register. July 2019. Rules and Regulations for the Classification of Ships, v8.1. https://www.lr.org/en-in/rules-and-regulations-for-the-classification-of-ships/.

Offshore Centre - Denmark. 2010. Offshore Book – An Introduction to the Offshore Industry. https://www.scribd.com/doc/125852163/Offshore-Book-2010.

The International Marine Contractors Association DP. Incident Reports. https://www.imca-int.com/divisions/marine/dynamic-positioning/dp-events-incidents.

The International Marine Contractors Association M 103. January 2019. Guidelines for the Design and Operation of Dynamically Positioned Vessels. https://www.imca-int.com/publications/57/guidelines-for-the-design-and-operation-of-dynamically-positioned-vessels/.

The International Marine Contractors Association M 103. December 2019. Guidelines for the Design and Operation of Dynamically Positioned Vessels. https://www.imca-int.com/publications/57/guidelines-for-the-design-and-operation-of-dynamically-positioned-vessels/.

The International Marine Contractors Association M 115. July 2016. Risk Analysis of Collision of Dynamically Positioned Support Vessels with Offshore Installations. https://www.imca-int.com/publications/77/risk-analysis-of-collision-of-dynamically-positioned-support-vessels-with-offshore-installations/.

The International Marine Contractors Association M 117. September 2016. The Training and Experience of Key DP Personnel. https://www.imca-int.com/publications/97/the-training-and-experience-of-key-dp-personnel/.

The International Marine Contractors Association M 125. September 2016. Safety Interface Document for a DP Vessel Working Near an Offshore Platform. https://www.imca-int.com/publications/83/safety-interface-document-for-a-dp-vessel-working-near-an-offshore-platform-includes-word-checklist/.

The International Marine Contractors Association M 131. September 1995. A Review of the Use of the Fan Beam Laser System for Dynamic Positioning. https://www.scribd.com/document/ 352592274/THE INTERNATIONAL MARINE CONTRACTORS ASSOCIATION - M-131-Review-of-the-Use-of-the-Fan-Beam-Laser-System-for-Dynamic-Positioning.

The International Marine Contractors Association M 140. January 2017. Specification for DP Capability Plots. https://www.imca-int.com/publications/111/specification-for-dp-capa bility-plots/.

The International Marine Contractors Association M 161. January 2019. Guidelines for the Design and Operation of Dynamically Positioned Vessels – Two-Vessel Operations – A supplement to The International Marine Contractors Association M 103. https://www. imca-int.com/publications/57/guidelines-for-the-design-and-operation-of- dynamically-positioned-vessels/.

The International Marine Contractors Association M 166. October 2019. Guidance on Failure Modes & Effects Analyses (FMEAs). https://www.imca-int.com/publications/179/gui dance-on-failure-modes-and-effects-analysis-fmea/.

The International Marine Contractors Association M 170. November 2003. A Review of the Use Marine Laser Positioning Systems. https://www.imca-int.com/publications/195/a-r eview-of-marine-laser-positioning-systems-part-1-mk-iv-fanbeam-and-part-2-cyscan/.

The International Marine Contractors Association M 242. January 2017. Satellite Positioning Systems. https://www.imca-int.com/publications/411/guidance-on-satellite-based-posit ioning-systems-for-offshore-applications/.

The International Marine Contractors Association. September 1993. Issue 2. Part 2. Guidelines for Auditing DP Vessels. https://www.scribd.com/document/246312741/ The International Marine Contractors Association m112-2-Guideline-for-Auditing- Vessel-With-DP-System.

International guidelines for the safe operation of dynamically positioned offshore supply vessels. https://www.imca-int.com/publications/236/international-guidelines-for-the-safe- operation-of-dynamically-positioned-offshore-supply-vessels/.

The International Marine Contractors Association. June 2011. Examples of a DP Vessel's Annual Trials Programme. https://www.imca-int.com/news/2013/01/21/conducting -annual-dp-trials-programmes-for-dp-vessels/.

Index